Business Lease Renewals: The New Law and Practice

Philip Freedman
in association with
Eric Shapiro and Kevin Steele

Routledge
Taylor & Francis Group
LONDON AND NEW YORK

First published 2006 by Estates Gazette

2 Park Square, Milton Park, Abingdon, Oxon OX14 4RN
711 Third Avenue, New York, NY 10017, USA

Routledge is an imprint of the Taylor & Francis Group, an informa business

First issued in hardback 2016

ISBN 13: 978-0-7282-0478-2 (pbk)
ISBN 13: 978-1-138-15601-2 (hbk)

Typeset in Palatino 10/12 by Amy Boyle, Rochester

Contents

Preface

The Landlord and Tenant Act 1954 remains the foundation of the law giving security of tenure relating to tenants of business premises, but has been radically amended with effect from 1 June 2004. This book sets out the new law but includes references to the old law wherever it may still be relevant.

The practice and procedure of lease termination and renewal is just as important as the law, and I thank Eric Shapiro for his valuable input from the surveyor's and valuer's viewpoint and Kevin Steele for providing the guidance on litigation procedures.

The law is stated as at 1 January 2006.

Philip Freedman
London

Table of Cases

The references are to paragraph numbers in the text

1 Overview of the Act

1.1 Legislation and regulations

The legislation on business lease renewal is principally contained in the Landlord and Tenant Act 1954, which in this book is called "the Act". The Act stood unamended for many years before it was found necessary to add to or change any of its major provisions. It is a testimony to the high quality of draftsmanship of the Act that it still forms the basis of the law of business lease renewals in England and Wales, none of the later amendments having radically changed the way in which the Act operates.

The first main amendments to the Act were made in 1969, when "contracting out" became permitted and the concept of "interim rent" was introduced. Then, in 1990, premises with liquor licences were brought within the Act. The next major changes were not made until 2003, taking effect on 1 June 2004, when some of the procedures and time-limits were changed and some other aspects of the legislation were revised. In between there have been a few minor amendments, mainly reflecting changes in other legislation referred to in the Act.

The first few sets of amendments to the Act were made by statutes, principally the Law of Property Act 1969 and the Landlord and Tenant (Licensed Premises) Act 1990. By contrast, the 2004 amendments were effected by a new type of Parliamentary instrument called a regulatory reform order, in this case the Regulatory Reform (Business Tenancies) (England and Wales) Order 2003, which came into force on 1 June 2004.

1.2 The 2004 amendments and notices

The changes made by the regulatory reform order were subject to transitional provisions under which they do not apply to anything consequent upon a section 25 notice or section 26 request given before 1 June 2004. Because there are still lease renewals in progress which were commenced by those types of notice given before that date, both the old and the amended provisions of the Act need to be covered in this edition of this book.

In this book , the "2004 amendments" means the amendments to the Act made as from 1 June 2004 by the Regulatory Reform (Business Tenancies) (England and Wales) Order 2003, and that order is referred to as "the RRO". References to "new" versions of sections of the Act are references to the sections introduced or amended by the RRO, and "old" versions are those prior to the 2004 amendments.

The Act requires certain notices to be in a form prescribed by regulations. The current forms for use in respect of property in England or in Wales are prescribed by the Landlord and Tenant Act 1954, Part 2 (Notices) Regulations 2004 ("the 2004 Notices Regulations"). The Welsh Assembly issued a consultation paper with a view to making an order prescribing, in respect of property which is wholly in Wales, versions of the same forms bilingual in Welsh and English, apart from the two forms which apply only to property in England. At the publication date of this book, the matter was still under consideration and it appeared likely that the use of bilingual forms would not commence before Autumn 2006.

1.3 Framework of the Act

The basic framework of Part II of the Act is that if a particular business tenancy has the protection of the Act then so long as the tenant remains in occupation of at least part of the premises carrying on a business, his tenancy will not automatically end on the expiry date of his lease. It can be ended only by following certain procedures, at which time the tenant is given the right to seek a new tenancy of those parts of the premises which he occupies (called his "holding"), at market rent, which the landlord can only refuse to grant if certain limited grounds apply. Further, compensation is payable to the tenant if the landlord successfully opposes the grant of a new tenancy on grounds other than those which involve the tenant having been in default of his obligations under his tenancy.

A tenant whose tenancy is protected by the Act (see Chapter 2) can therefore remain in possession of the premises let to him, beyond the expiry of his contractual tenancy, and his tenancy will continue (as explained in Chapter 4) until it is ended by one of the procedures mentioned in the Act (set out in Chapters 4 to 7). The grounds on which the landlord might oppose the grant to the tenant of a new lease on the termination of his tenancy are explained in Chapter 8.

Certain types of business tenancy are excluded from the protection of the Act. These are set out in Chapter 2. The main method of excluding protection when granting a tenancy is described in Chapter 3.

1.4 Procedures under the Act

The Act imposes several procedural steps to be taken by landlords and tenants. A failure to understand and observe the procedures can be highly detrimental; the landlord who fails to follow the statutory process may suffer delays in securing an increase in rent or in obtaining possession; the tenant who does so might lose his protection. Acquiring familiarity with the operation of the Act is essential for professionals advising owners and occupiers about lease termination and renewal.

These essential procedures are explained fully in Chapters 5 to 10. Essentially, if the landlord gives the tenant a termination notice under the Act but the tenant wishes to remain in possession beyond the expiry of that notice, he must either reach agreement with the landlord for a new tenancy (as explained in Chapter 13) or ensure that an application to the court is made before the specified termination date unless an extension of that date is agreed in writing with the landlord for this purpose (discussed in Chapter 10).

The special considerations in relation to premises which the tenant has sublet are discussed in Chapter 9.

If he complies with the statutory procedures, the tenant will be entitled to be granted a new tenancy on his holding, unless the landlord successfully opposes the grant of a new tenancy on one or more of the grounds listed in the Act. These grounds are explained in Chapter 8.

If the landlord does not oppose the grant of a new tenancy, or attempts to oppose it but fails to establish one or more of the grounds listed in the Act, the tenancy to which the tenant is entitled to be granted is to be on terms to be agreed between the tenant and the landlord. If agreement on any aspect cannot be reached, the disputed terms will be fixed by the court in accordance with guidelines set out in the Act, as described in Chapter 12. The court process itself is described in Chapter 10.

It is possible for the parties to agree that they will refer one or more matters to an arbitrator or expert instead of the court, but certain court processes may still be necessary, as explained in Chapter 10.

If the landlord successfully opposes the grant of a new tenancy on one or more of the grounds under the Act, the tenant must vacate the property by a date ascertained under the provisions of the Act, considered in Chapter 10. In some cases the tenant will be entitled to claim compensation from the landlord, as set out in Chapters 14 and 15.

If the tenant does not want to claim a new tenancy, he can vacate by the end date of his lease as discussed in Chapter 3. He can, but need not, give three months' notice to the landlord. He may do this either voluntarily or when he receives a termination notice under the Act. If he simply vacates, the landlord may not know that the tenancy will not be continued.

Even if he applies for a new tenancy, the tenant can change his mind. The Act permits him to give three months' notice to end his tenancy at any time if he is simply holding over after the expiry of his lease. If an application to the court is made regarding a new tenancy, the tenant may at any time stop the court application and his tenancy will end three months later. Even if the court process is followed and the court makes an order for the granting of a new tenancy, the tenant may have the order cancelled within 14 days of its being made, with the tenancy ending three months later. In some of these cases the tenant may be exposed to a claim for costs from the landlord.

Interim rent, which may be payable by the tenant for the period while a tenancy is being temporarily continued under the Act, is considered in Chapter 11.

1.5 Importance of defined words and phrases

A glossary of words and phrases appears towards the end of this book. The Act contains many defined terms which need to be understood in order to follow the way in which the Act operates. For example, that part of the premises let to the tenant which the tenant actually occupies is called the "holding"; the person who is the "landlord" may, in certain circumstances which the Act specifies, be in fact a superior landlord rather than the tenant's immediate landlord; and it is necessary to understand the distinction between the contractual term expressly created by a lease or tenancy agreement and the "tenancy" that it creates.

The glossary sets out the most often used defined words and phrases and their meanings. These special meanings should always be remembered whenever those words or phrases appear.

As stated above, in this book "the Act" means the Landlord and Tenant Act 1954; all references to sections and schedules without mentioning the particular statute are references to sections or schedules of the Act.

1.6 Use of casenotes

Short summaries of reported court cases have been included in certain sections of the book, to illustrate some of the more complicated issues that arise under the Act.

There have been literally hundreds of reported cases on the Act and choosing those to summarise here has involved a difficult selection process. Generally, the casenotes concentrate on decisions of the Court of Appeal or House of Lords, since these are authoritative as precedent in English law. Decisions in the High Court at first instance, and decisions of county courts, are not generally binding precedent and are only cited where they carry the weight of many years authority or illustrate a novel point.

New cases on the Act are reported every few months and no legal book can remain up-to-date on every point for very long. If a serious issue arises under a particular provision of the Act, it is advisable to obtain specialist legal advice and for the legal adviser to carry out research into all the relevant case law. The *Estates Gazette*, including its EGi on-line service, is a good source of specialist property law reporting, in addition to the official law reports and those of other legal publishers.

1.7 Stamp duty land tax

The chapters dealing with continuation tenancies, interim rent and the grant of the new lease include some brief guidance on the tenant's potential liability to pay stamp duty land tax (SDLT).

SDLT is a new tax on property transactions and it replaced stamp duty as from 1 December 2003. It is still in its infancy and subject to frequent amendment, so several aspects of its operation are somewhat uncertain and the guidance set out in this book is inevitably tentative in some respects. Specialist advice should be taken whenever this may be an issue.

Tenancies to Which the Act Applies

2

2.1 Overview of main rules

For the protection given by the Act to apply, the criteria contained in section 23 must be present. In that section, subsection (1) states:

> Subject to the provisions of this Act, this Part of this Act applies to any tenancy where the property comprised in the tenancy is or includes premises which are occupied by the tenant and are so occupied for the purposes of a business carried on by him or for those and other purposes.

This has four elements. There must be a "tenancy". The tenancy must be of "premises". At least part of the premises must be "occupied" by the tenant. Finally, the occupation must be "for the purposes of a business". These criteria are each described in more detail below.

If the parties litigate the issue of renewal under the Act, the tenant must demonstrate that he has met these criteria continuously throughout the period of the proceedings up to and including the date of the court hearing (*Domer* v *Gulf Oil (GB) Ltd* (1979) 119 SJ 392, Ch D).

The question whether an arrangement constitutes a "tenancy" (having the protection of the Act) or a mere licence or tenancy-at-will (which would be unprotected) turns on the individual facts of each case. While the distinctions are discussed below, it is dangerous to try to avoid the Act by purporting to grant a licence where, on the facts of the arrangement, the law is likely to view the transaction as really the grant of a tenancy; there are simpler ways of excluding tenancies from the protection of the Act which give more certainty, for example, by a validated agreement excluding sections 24 to 28 as discussed below.

The protection given by the Act will not apply if one of the following exclusions is present:

(a) there is a term of the tenancy prohibiting any form of business use, and that term has not been waived or a breach acquiesced in or
(b) there is a provision of the tenancy stating that the parties have agreed to exclude sections 24 to 28, and that provision was validated by following the procedure referred to in section 38 of the Act or
(c) the tenancy is for a fixed term not exceeding six months with no right for the tenant to renew or extend it, and the tenant has not been in occupation for 12 months or more or
(d) the tenancy is merely a tenancy-at-will or
(e) the premises fall into certain exempt categories or

(f) the tenant is an office-holder or employee of the landlord and the tenancy will end when the office or employment ends.

These exclusions will be described in more detail.

In some situations, generally where the landlord is a government department, local authority or similar public body, the protection of the Act may be restricted or adapted, for example by excluding any right of the tenant to apply to the court for a new tenancy or by restricting the terms of the new tenancy. Since these are effected by altering the notice procedures and new tenancy terms under the Act, they are dealt with in Chapters 6 and 12.

2.2 Tenancies as distinct from licences

For an occupier to have the protection of the 1954 Act, he must be a "tenant", which means that he must hold the premises under a tenancy. A "tenancy" is defined in section 69(1) of the Act as:

> a tenancy created either immediately or derivatively out of the freehold, whether by a lease or underlease, by an agreement for a lease or underlease or by a tenancy agreement ...

A tenancy which is purported to be granted by a person who does not in fact have any legal interest in the property can be a "tenancy" within section 69(1) as between the parties to the tenancy, because of the legal doctrine of estoppel, such as where the tenancy is mistakenly granted by an associated company of the freeholder. In such cases the court might also treat the purported landlord as the owner of the freehold by estoppel (*Bell* v *General Accident Fire & Life Assurance Corporation Ltd* [1998] 1 EGLR 69, CA).

Protection is not given to occupiers who occupy premises under an arrangement which, in law, is less than a tenancy. Examples of tenancies are leases and tenancy agreements. Examples of lesser types of occupancy rights are "licences" or "tenancies-at-will".

A tenancy will generally exist if the substance of the arrangement between the owner of the property and the occupier is that the owner gives the occupier exclusive possession of the premises for a continuous period of time, particularly if in return for rent (*Street* v *Mountford* [1985] 1 EGLR 128). Even if occupation is given for a continuous term at a rent, there will be no tenancy unless the right to exclusive possession is given (*National Car Parks Ltd* v *Trinity Development Co (Banbury) Ltd* [2001] PLSCS 222, CA).

This arrangement is most commonly created by a document called a lease or tenancy agreement. The period of exclusive possession that it grants to the occupier is called the term. A lease is a document which lets the premises for a "term of years absolute" (Law of Property Act 1925, section 1(1)(b)), which means a term of fixed duration although it may contain provisions for the period to be ended earlier or to continue after the end of the fixed period. A tenancy agreement is usually a document which lets the premises on a periodic basis, such as monthly or quarterly.

With one exception, a lease has to be in writing and must be signed by the parties as a deed (Law of Property 1925, sections 52, 53). The exception is that a written document is not needed if the term does not exceed three years, provided that the term starts immediately and the rent is the best rent reasonably obtainable. If the term granted by a lease is for over seven years, the lease must be registered at the Land Registry (Land Registration) Act 2002, section 4(1)(c)(i)); before 13 October 2003, only leases for terms exceeding 21 years needed to be registered. Under these rules, a tenancy which is periodic (such as weekly, monthly or yearly) need not be created in writing and can arise out of the dealings between the parties. For example, if a shopkeeper rents a lock up garage at a monthly rent, without any written document, that may create a monthly tenancy.

Rent is not always essential for the existence of a tenancy. Where exclusive possession of business premises is given for a period of time, there may in some cases be a tenancy even if the arrangement is rent free.

If the premises are shared rather than exclusively possessed, or are in a non-fixed location (such as a moveable franchise counter in a department store), or are granted only for temporary use without any intention to create the type of relationship which the law considers to be a tenancy, then a "licence", rather than a tenancy, will normally exist and the occupier will not have the protection of the Act. Unfortunately, in its application to particular facts, the distinction between "tenancy" and "licence" can often be indistinct and uncertain. Even if there is a document governing the arrangement and it describes itself as a "licence", the occupier may claim that it amounts to a tenancy and then the court will consider the substance of the arrangement and the intention of the parties as is derived from the document and from their conduct (*Addiscombe Garden Estates Ltd* v *Crabbe* [1958] 1 QB 513).

Examples of a genuine licence might be where the owner of premises (who might be the freeholder or a tenant) permits a business associate to occupy a desk within his office, even if he makes a charge for this; where a retail store operator grants a branded goods franchise to be conducted at a counter in the store; where a petrol company agrees that an operator will run a petrol filling station from one of its sites; or where a right is granted, even on an exclusive basis, to dump waste on land.

Advertising sites, such as the visible flank walls of buildings, are frequently the subject of arrangements with advertising companies using documents which describe themselves as licences. However, if exclusive possession is given and the site can properly be described as comprising "premises" (see 2.5), then a tenancy may exist.

Casenotes

Shell-Mex and BP Ltd v *Manchester Garage Ltd* [1971] 1 WLR 612, CA: An oil company granted a "licence" of premises for one year for use as a petrol filling station selling oil and petrol supplied only by the oil company. The document stated that the oil company retained possession and control of the property and could alter and decorate it. At the end of the licence period, the "licensee" claimed the protection of the Act and refused to vacate. Held: This arrangement was a genuine licence and not a tenancy and so the occupier was not protected by the Act. (For a more recent case where a similar document had been used, see *Esso Petroleum Co Ltd* v *Fumegrange Ltd* [1994] 2 EGLR 90).

Street v *Mountford* [1985] 1 EGLR 128: The top floor of a house was occupied under a "licence agreement" which contained a statement that the licensee understood that she was not being given a tenancy. Held: Since exclusive possession was given to the occupier for a period in return for payment, it was a tenancy notwithstanding that it purported to be only a licence (being residential, it was protected by the Rent Acts).

AG Securities v *Vaughan* [1988] 2 EGLR 78: Separate licences were given to individuals to share a flat. Held: Since exclusive possession was not given to anyone, each of the sharers had a mere licence and not a tenancy.

Dresden Estates Ltd v *Collinson* [1987] 1 EGLR 45: An express licence was granted in which the "licensor" reserved the right to move the "licensee" from one part of the licensor's premises to another. Held: Since the licensee was not being given exclusive possession of any specific premises, the arrangement was a mere licence and not a tenancy.

Manchester City Council v *National Car Parks Ltd* [1982] 1 EGLR 94: A company which intended to develop some land in the future granted NCP a "licence" to use the land as a car park, but only from midnight to 2 am and from 7 am to midnight, daily, for six months but terminable on short notice for redevelopment. There were good commercial reasons for imposing the hours limits and the other terms. Held: This arrangement was a licence and not a tenancy.

National Car Parks Ltd v *Trinity Development Co (Banbury) Ltd* [2001] PLSCS 222, CA: The owner of a car park granted NCP a "licence for the parking of vehicles" from year to year until ended by either party on three month's notice. It provided that NCP was to operate the car park but had to ensure that 40 spaces were available for use by the occupiers of the licensor's adjacent property. NCP was to "manage and administer" the car park but must "not impede in any way the officers ... of the Licensor in exercise of the Licensor's rights of possession...". NPC was to pay the owner a percentage of its net profit from the car park. The owner conceded that occupation had been granted for a term at a rent, but denied that exclusive possession had been give. Held: Exclusive possession had not been given. This was a licence, not a tenancy, and it was not protected by the Act.

Clear Channel UK Ltd v *Manchester City Council* [2005] CA Civ Div 9 Nov: The council reached an initial agreement with Clear Channel that, at various sites which were listed, Clear Channel would construct and operate certain M-shaped advertising stations placed on concrete bases and would pay rent for this. An agreement was drafted but never signed, although the arrangements were implemented and rent was paid in accordance with the draft agreement. The agreed draft agreement stated the addresses of the council's land in the various locations but did not identify exactly where the advertising stations were to be erected on them, although the agreed locations for the stations were subsequently marked with spray paint on the ground. The council told Clear Channel that it was terminating the arrangement but Clear Channel claimed that it had the protection of the Act in respect of each of the advertising stations. The council then reached an agreement with Clear Channel in respect of a particular site and produced a draft agreement which showed an intention to let to Clear Channel the particular area of land on which the concrete base was to be built at that site. Again the agreement was never signed but the construction was carried out and the rent was paid. Held on appeal: Since the initial arrangement did not specify that a particular area at any of the listed sites would be permanently allocated to the advertising stations, or identify any such part of the land, there was no intention to grant exclusive possession and thus no tenancy. (At first instance it had been held, in relation to one site, that the subsequent separate arrangement for that one site did indicate an intention to give exclusive possession of an agreed part of that particular site and its implementation created an annual tenancy of the land on which the concrete base was built at that site. That aspect was not the subject of the appeal.)

2.3 Tenancies-at-will

A "tenancy-at-will" is outside the protection of the Act since it is not a "tenancy" for the purpose of section 23. A tenancy-at-will is an occupancy arrangement which can be ended by either party "at will", that is by giving immediate notice of termination at any time.

A tenancy-at-will is usually found to exist by implication when an unprotected tenant — such as a new occupier, or an existing tenant whose tenancy had been excluded from the Act or who had lost his right to apply to the court for a new tenancy — is allowed to take or remain in occupation while negotiations are taking place for granting to him a new tenancy, it being the intention of the parties that no tenancy is being created by that conduct during the negotiating period (*Cardiothoracic Institute* v *Shrewdcrest* [1986] 2 EGLR 57; *Javid* v *Aqil* [1990] 2 EGLR 82 CA). However, if occupation is allowed to continue on a rent-paying basis without negotiations continuing and without the parties having agreed the status of that arrangement, it may be considered to create a periodic tenancy protected by the Act (*Walji* v *Mount Cook Land Ltd* [2001] PLSCS 2, CA).

The courts will be cynical when considering the genuineness of any written tenancy agreement which is expressed to be a tenancy-at-will. If the court perceives that there is an understanding between the parties that the occupancy will not in fact be ended by immediate notice at the whim of either party, the court may decide that it is not truly "at will" but is a proper tenancy protected by the Act.

If a tenant who is protected by the Act enters into a new occupancy arrangement with the landlord which amounts to a tenancy at will, he will be treated as having surrendered his previous tenancy and will lose the protection of the Act (*Gibbs Mew plc* v *Gemell* [1999] 1 EGLR 43, CA).

Casenotes

Wheeler v *Mercer* [1957] AC 416, HL: The landlord of premises gave the tenant notice to quit before the Act came into force in 1954. The tenant remained in possession while he negotiated with the landlord for a new lease. The Act then came into force and the tenant claimed its protection on the basis that he still had a tenancy of the premises. Held: The House of Lords ruled that the Act does not apply to a tenancy-at-will, and the tenant's continued occupation in these circumstances was under a tenancy-at-will since either party could have ended the negotiations at any time, in which case the tenant would have been expected to vacate.

Hagee (London) Ltd v *AB Erikson & Larson* [1976] QB 209, CA: Hagee held the premises under a lease which prohibited subletting. In 1971 they agreed with the defendants, who wanted to be given a sublease, to let them occupy part of the premises "as tenants at will only" at an annual rent of £5000 payable quarterly in advance, with provision that they would on termination refund any payment made in advance. Hagee gave the occupiers a termination notice in 1973 but they refused to vacate and claimed the protection of the Act. Held: On the facts, a mere tenancy at will had been created; however Lord Denning MR and Scrutton LJ warned that the court will look closely in these types of case to see whether or not there really is a tenancy-at-will or whether the document actually creates a periodic tenancy.

Cardiothoracic Institute v *Shrewdcrest* [1986] 2 EGLR 57: The premises had been let to the tenant under a series of short leases, each being excluded from the Act by court orders under section 38(4), since the landlord wanted the freedom to redevelop the property at some future date. The last such lease expired in October 1983. Negotiations for a further short excluded lease started before that date but remained unconcluded, the tenant staying in occupation and paying agreed monthly payments of rent; this continued for just under two years. The tenant then claimed to be protected by the Act. Held: Both parties intended that there would be no legally binding new tenancy until the terms had been agreed and an exclusion order had been obtained from the court. The continued possession and payment of rent in the meantime did not create a periodic tenancy.

Javid v *Aqil* [1990] 2 EGLR 82 CA: The occupier was let into possession on paying £2500 "as rent for three months in advance", it being envisaged that a formal lease would be negotiated. Those negotiations were almost concluded and two further payments of rent were made. Ultimately the negotiations foundered, but the prospective tenant refused to vacate and claimed protection of the Act. Held: The parties had had no intention to create a proper tenancy until a formal lease was completed, so this arrangement was merely an unprotected tenancy at will.

Walji v *Mount Cook Land Ltd* [2001] PLSCS 2, CA : The tenant company, controlled by the claimants, was struck off the register of companies. The landlord agreed, on a subject to contract basis, to grant the claimants a new lease. Over a period of over three years they and their successors accepted the agreed quarterly rent from the claimants and allowed them to occupy the premises but did not press for completion of the new lease. Nothing had been said about the status of the claimants pending completion of the lease. Held: The claimants had a periodic tenancy protected by the Act, not a tenancy at will, since terms had been agreed and there were no continuing negotiations.

2.4 Subtenancies

The expression "tenancy" includes a subtenancy (section 69). Subtenants can have the protection of the Act both as against their immediate landlord and the superior landlord (*HL Bolton (Engineering) Co Ltd* v *TJ Graham & Sons* [1965] 1 QB 159, CA). The procedure for dealing with subtenants is explained in Chapter 9.

This protection even extends to "unlawful" subtenants, where the sublease was granted in breach of a tenant's covenant not to sublet, or where the tenant did not obtain the landlord's consent to sublet where the terms of his tenancy required him to do so.

Casenote

D'Silva v *Lister House Development Ltd* [1971] Ch 17: The tenancy of a medical suite prohibited sublettings without the landlord's consent. The tenant agreed terms of subletting with a foreign doctor, let him into possession, accepted rent, and then sought the landlord's consent for the grant of a formal sublease; consent was refused. The foreign doctor refused to vacate when the tenant's own lease ended, and he claimed the protection of the Act. Held: The arrangement between the tenant and the doctor amounted to a subletting. Although it was an unlawful subletting, being in breach of the tenant's lease, the subtenant had the protection of the Act against the superior landlord at the end of his immediate landlord's tenancy.

2.5 What constitutes "premises"

For protection to arise under the Act, the tenancy must comprise "premises" capable of being occupied. The word "premises" for this purposes is not limited to buildings or even land with buildings, but covers any kind of land in respect of which a lease is granted (*Bracey* v *Read* [1963] Ch 88).

Usually this requirement for "premises" does not pose a problem, but this is not invariably the case and it may be unclear whether, in any particular factual situation, what is expressed to be let comprises "premises" for this purpose. Examples are leases of easements such as rights of way, and lettings of fixtures such as advertising hoardings. However, where rights of way or other rights over adjacent land are included in a lease of land or buildings which are occupied for business purposes and the rights are not expressed to be personal to the tenant, those rights will be included within the protection of the Act and any renewal of the lease will include a renewal of those rights (*Nevill Long & Co (Boards) Ltd* v *Firmenich & Co* (1983) 2 EGLR 76, CA).

The requirement for the existence of "premises" is sometimes related to the issue whether the relationship between the parties constitutes a "tenancy". For example, the question whether an agreement granting rights for the erection and display of an advertisement hoarding or other structure falls within the Act may turn upon whether the grantee has being granted "premises" and whether exclusive possession has been given to him in respect of those premises (see 2.2).

Casenotes

Bracey v *Read* [1963] Ch 88: The owner of gallops let a racehorse trainer into possession of the gallops for training and exercising racehorses, at a rent. Held: "Premises" in section 23 was not confined to buildings but included open land. The trainer had the protection of the Act.

Re Pitsea Access Road, Land Reclamation Co v *Basildon District Council* [1979] 1 WLR 767: A lease was granted of a right of way over a road. Held: This was not a lease of the road itself but merely of a right over it; therefore it was not a tenancy of "premises".

2.6 Occupation by the tenant

For a tenant to have the protection of the Act at any point in time, he must then be occupying at least some part of the premises comprised in his tenancy. A tenant can occupy either personally or through the presence on the premises of his employees. The purpose of this occupation must be, at least in part, for the tenant's business, which will be discussed next.

In particular, in order to be protected by the Act the tenant must be in occupation, and in those cases where the landlord gives a notice to quit, this must be the case at the date on which that notice is given;

occupying only after that date does not prevent the notice from ending the tenancy (section 24(3)(a)). In cases where the tenant applies to the court to order a new tenancy, he must also be in occupation at the date of that application. There are however a few exceptions to these rules, described below.

The requirement for occupation does not literally mean that the tenant cannot leave the premises unused for occasional periods, so long as no other person is occupying the premises as a tenant during that period and the use of the premises has not been changed to a non-business use (*I&H Caplan Ltd* v *Caplan (No 2)* [1963] 1 WLR 1247; *Bacchiocchi* v *Academic Agency Ltd* [1998] 3 EGLR 157).

Provided that the tenancy is for a continuous term, the fact that the business is conducted from the premises only seasonally does not prevent the tenant from being treated as in occupation throughout the year (*Artemiou* v *Procopiou* [1966] 1 QB 878, CA).

If the tenant has been occupying the premises for the purpose of his business but that occupation is interrupted by matters beyond his control, eg fire damage, he will be taken to be continuing in occupation so long as he evinces an intention to resume occupation as soon as that becomes possible.

Casenotes

Wandsworth London Borough Council v *Singh* [1991] 2 EGLR 75: A piece of land was let to a local council for use as an open space. The council through their parks staff controlled its use and maintained it. Held: It was "occupied" by the council.

Latif v *Hillside Estates (Swansea) Ltd* [1992] EGCS 75, CA: Premises were not occupied by the tenant for business purposes until five months after he made his court application. Held: His tenancy was unprotected.

Artemiou v *Procopiou* [1966] 1 QB 878, CA: A seaside café and restaurant was let on a two year lease. The café was not operated out of the summer season. Held: Occupation during the summer season only was sufficient occupation for protection under the Act.

Morrison Holdings Ltd v *Manders Property (Wolverhampton) Ltd* [1976] 1 EGLR 70: The tenant had to vacate the premises because they were substantially damaged by fire. The tenant urged the landlord to reinstate so that he could resume business in the premises. The landlord served a break notice on the tenant under a provision in the lease, but did not serve a notice under the Act. Held: The tenant was treated as still being in occupation, and thus protected by the Act, since he had evinced an intention to resume occupation and he had been dispossessed due to reasons beyond his control.

2.7 Occupation by a group company

If the tenant is a "body corporate", occupation by another body corporate in the same "group" counts as occupation by the tenant for the purpose of meeting the requirement that the tenant must be in occupation (section 42(2)).

There are three situations in which two bodies corporate are treated as members of the same "group". The first is where one is a "subsidiary" of the other. The second is where both are subsidiaries of a third body corporate. The third, which applies only to cases covered by the 2004 amendments (see 1.2), is where both are controlled by the same individual (section 42(1)).

It may be anomalous but it may be that the first two types of "group" extend to a wider range of bodies corporate than the third. The reason is that section 42, although headed "Groups of companies", does not use the word "company" but rather "body corporate" and it refers to "subsidiary" within the definition in section 736 of the Companies Act 1985 Act. Section 736 of the 1985 Act provides:

(1) A company is a "subsidiary" of another company, its "holding company", if that other company —

> (a) holds a majority of the voting rights in it, or
> (b) is a member of it and has the right to appoint or remove a majority of its board of directors, or
> (c) is a member of it and controls alone, pursuant to an agreement with other shareholders or members, a majority of the voting rights in it,
>
> or if it is a subsidiary of a company which is itself a subsidiary of that other company.
>
> (2) A company is a "wholly-owned subsidiary" of another company if it has no members except that other and that other's wholly-owned subsidiaries or persons acting on behalf of that other or its wholly-owned subsidiaries.
>
> (3) In this section "company" includes any body corporate.

Having regard to the final subsection set out above, the first two types of "group" covered by section 42 apply to groups which comprise or include not only companies registered under the Companies Acts but also other types of "body corporate" such as charter bodies, limited liability partnerships and foreign companies.

This is to be contrasted with the third type of group which comprises two "bodies corporate" companies where the same individual has a "controlling interest" in both of them. Section 46(2) of the 1954 Act provides that:

> a person has a controlling interest in a company if, had he been a company, the other company would have been its subsidiary; and ... "company" has the meaning given by section 735 of the Companies Act 1985; and "subsidiary" has the meaning given by section 736 of that Act.

The Act does not contain any definition of "controlling interest" in relation to any body corporate other than a company within the meaning of section 735 of the 1985 Act. That meaning is restricted to companies which are or were registered under the Companies Acts. For this reason it is arguable that the third type of "group" covered by section 42 is restricted to the case where the same individual controls two or more companies which have been registered under the Companies Acts (as defined in section 735), but does not apply where he controls any other type of body corporate such as a foreign company or a limited liability partnership. This is consistent with the other provisions about companies controlled by an individual, mentioned next, but is inconsistent with the other "group" provisions of section 42.

2.8 Occupation by a company controlled by the tenant and vice versa

Before the 2004 amendments, if the tenant was an individual, the fact that he controlled the company which was in occupation did not mean that he was treated as being in occupation through that company, even if he was personally running the company's business in the premises; in that situation there was no protection (*Cristina* v *Seear* [1985] 2 EGLR 128; *Nozari-Zadeh* v *Pearl Assurance plc* [1987] 2 EGLR 91).

Where the 2004 amendments apply (see 1.2), occupation by a company in which the tenant has a controlling interest now counts as occupation by the tenant himself, as does occupation, where the tenant is a company, by a person with a controlling interest in the tenant company (section 23(1A)). This means that there is now protection under the Act where, for example, a business partnership holds the tenancy of its premises through a service company controlled by the partners, or persons who were in partnership incorporate a company to carry on their business but continue holding the tenancy in their personal names.

This is a major change which allows the individuals who hold the lease to protect the lease from the business risks affecting their trading company — if the company becomes insolvent, the company's liquidator may have no claim on the lease and the liquidation of the company will not normally entitle the landlord to seek forfeiture of the lease since the company is not the tenant.

Section 46(2) of the 1954 Act provides that:

> a person has a controlling interest in a company if, had he been a company, the other company would have been its subsidiary; and ... "company" has the meaning given by section 735 of the Companies Act 1985; and "subsidiary" has the meaning given by section 736 of that Act.

Since "company" as defined in section 735 of the Companies Act 1985 is restricted to companies registered under the Companies Acts (as defined in section 735), these provisions do not include other types of corporate bodies such as foreign companies or limited liability partnerships. This should be contrasted with certain of the group provisions of section 42 (see 2.7). If a tenancy is held by persons who are members of a limited liability partnership, the occupation of the premises by the limited liability partnership for its business will not count as occupation by those persons and there will be no protection under the Act unless they can demonstrate the existence of the trust situation discussed below.

2.9 Occupation by partnerships

Where the tenancy is held in the joint names of two or more persons in partnership (whether or not the partnership also includes other partners who are not named as tenants) as partnership property, special provisions of the Act deal with the common situation of partners leaving and joining the partnership.

If one or more named tenant leaves the partnership and vacates the premises, leaving one or more of the other named tenants remaining in the premises, the protection of the Act is not lost; those tenants who remain carrying on business in the premises (whom the Act calls the "business tenants") have the right to serve notices on the landlord under sections 26 and 27 (see 4.6, 4.7 and 7.1) without involving the partner or partners who have left, whom the Act calls "the other joint tenants" (section 41A).

In some cases it may be necessary to decide if it can be inferred that the lease is held on trust for a partnership, for example where the partner who holds the lease of the partnership premises dies and the lease vests in his executor (*Hodson* v *Cashmore* (1972) 226 EG 1203). In such cases where the trustee is not himself a partner, the trust rules set out next may apply instead of the partnership rules.

2.10 Occupation by beneficiaries of trusts

Where the tenants are trustees of a trust, the occupation of the premises by some or all beneficiaries of the trust and the carrying on there of a business by the beneficiaries counts for the purposes of section 23 as occupation by the tenants so that the tenants will be protected by the Act (section 41). However, for protection to apply to the trustees under this provision, the occupation by the beneficiary must be by virtue of his being a beneficiary and not, say, by being granted a sub-tenancy by the trustees (*Frish* v *Barclays Bank* [1955] 2 QB 541.

Any change in the persons who are the trustees is not treated as a change of tenant for the purpose of the Act (section 41(1)(c)).

2.11 Management agreements

Two separate issues arise in relation to management agreements. The first is whether premises can be treated as "occupied by the tenant", so as to give the tenant protection under the Act, if they are occupied by someone who is employed to manage the business at the premises on behalf of the tenant. The second issue is whether, as between the employer (whether he is a freeholder or a tenant) and his manager, the manager is to be treated as having his own tenancy and occupying the premises for his own business, in which event he may have the protection of the Act against his "employer".

On the first issue, premises can be "occupied by the tenant" by virtue of being occupied by the tenant's employees in the tenant's business which is carried on there. The definition of the "holding" in section 23(3) refers to parts of the premises being occupied by the tenant or "a person employed by the tenant and so employed for the purposes of a business by reason of which the tenancy is one to which this Part of this Act applies". The position is therefore clear if the employees are employed directly by the tenant and work in the tenant's business. The position becomes less clear if the business is directed by an agent or manager, especially where the manager employs the staff at the premises or has a personal stake in the business beyond that of a mere employee. Where the manager is genuinely running the business for the tenant, the tenant may be treated as being in occupation and protected by the Act, but the court will look at the full facts to see if in fact the so-called manager is running his own business in the premises and not the business of the tenant (*Teesdale* v *Walker* (1958) 172 EG 297).

On the second issue, as to the status of the manager's occupation, there are three possible situations. Where the manager is merely an employee of the owner of the premises, his presence at the premises will normally be in that capacity and, even if he were treated as having a tenancy, he will usually be excluded from the ambit of the Act by section 43(2). This is discussed at 2.19. Where instead there is a management agreement under which a person or company is engaged to run the business for the owner, that will usually amount to the grant to the manager of a mere licence to be in the premises and so, not being a tenancy, will also be outside the scope of the Act (as discussed earlier). However, if the arrangements go beyond mere management on behalf of the owner, such as where the so-called manager is in reality being given the right to trade on his own account from the premises, that arrangement may fall within the protection of the Act by being considered to create a tenancy (*Wang* v *Wei* [1976] 1 EGLR 66, Ch D; *Dellneed* v *Chin* [1987] 1 EGLR 75).

Casenotes

Wang v *Wei* [1976] 1 EGLR 66, ChD: The owner of a restaurant became ill and entered into a "management agreement" with Mr Wang. This gave him the right to run the restaurant and provided for him to inject capital into the business, and both parties treated the arrangement as if it was a tenancy. Held: Mr Wang had been given a tenancy protected by the Act.

Dellneed v *Chin* [1987] 1 EGLR 75: A restaurateur granted a company a "Mai Toi" management agreement to run his restaurant. Under the agreement, a young inexperienced man was allowed to run the restaurant on his own account, albeit under supervision. The agreement, although expressly excluding the creation of a tenancy, effectively gave the young man exclusive possession and imposed tenant-like duties and restrictions on him. Held: The arrangement gave the young man a tenancy protected by the Act.

2.12 Tenants who have sublet

Where the tenant sublets part of the premises demised to him, he will not be treated as being in occupation of the parts he has sublet (*Graysim Holdings Ltd* v *P&O Property Holdings Ltd* [1996] 1 EGLR 109, HL). If the tenant continues to occupy other parts for the purposes of his business, he may retain the protection of the Act due to his occupation of those retained parts. In that situation, the continuation provisions of the Act will apply to the whole premises demised to the tenant (including the sublet parts) but the tenant's right to seek a new tenancy may only apply to those parts (the "holding") which he occupies — this is more fully explained in Chapters 4 to 7 — and the subtenants may have their own rights of continuation and renewal of their subtenancies.

Where all the lettable parts of the premises have been sublet, usually the tenant will be treated as not occupying any part of the premises himself and he will therefore not retain the protection of the Act for his tenancy; this will apply even if technically he has retained the possession of certain parts of the premises which are ancillary to the lettable parts, as in the case of common areas, boiler rooms, porters' offices and similar ancillary accommodation (*Bagettes Ltd* v *GP Estates Ltd* [1956] 1 Ch 290; *Graysim Holdings Ltd* v *P&O Property Holdings Ltd* [1996] 1 EGLR 109, HL; *Bassari* v *Camden London Borough Council* [1998] EGCS 27, CA). The reasoning is that those ancillary areas are not capable of being a separate business holding.

There is a possible exception where the tenant carries on the business of letting serviced flats in the premises and maintains active control of the whole premises and a degree of presence there, particularly where he retains keys to the individual flats and exercises rights of entry. This might possibly amount to sufficient occupation by him of the flats as well as the common parts to enable his tenancy of the whole premises to be protected by the Act (*Lee-Verhulst (Investments) Ltd* v *Harwood Trust* [1973] QB 204), but this may be exceptional and does not easily sit with the more recent decision of the House of Lords in *Graysim*.

Casenotes

Bagettes Ltd v *GP Estates Ltd* [1956] 1 Ch 290: The tenant of a building sublet the individual flats but retained control and management of the common parts. Held: The tenant did not occupy the flats which he had sublet, and the common parts were used by the subtenants as ancillary to their flats, so the tenant did not occupy any part of the building for the purposes of the 1954 Act.

Lee-Verhulst (Investments) Ltd v *Harwood Trust* [1973] QB 204: The tenant company sublet most of its building as furnished flats. The owner of the tenant company lived in the building, retained pass keys and provided room service in each of the rooms. Held: The tenant company occupied the building for the purpose of the business of providing furnished flats and was protected by the Act. (However, see the commentary above).

Ross Auto Wash Ltd v *Herbert* [1979] 1 EGLR 95, Ch D: The tenant of the ground floor of a building carried on the business of granting trade concessions by way of licences to persons who sold goods from mobile stalls within the building. The tenant managed the premises and provided services to the licensees, including running their stalls for short periods in a day in the absence of the licensee. Held: The tenant was in occupation and was protected by the Act.

Hancock & Willis v *GMS Syndicate Ltd* [1983] 1 EGLR 70, CA: Tenants of premises granted someone a six months' "licence" giving him exclusive occupation of the premises subject to retaining the right to use part as a wine cellar themselves and to hold staff lunches at the premises twice a week. Held: The tenants had not retained sufficient use of the premises for them to be treated as occupying the premises. The tenants were unprotected.

Groveside Properties Ltd v *Westminster Medical School* (1984) 47 P&CR 507: The medical school had a tenancy of a flat. It granted four of its students a licence to live in the flat. The school retained the keys and the school secretary regularly visited the flat and spent time there. Held: The school occupied the flat for the purpose of its school business and its tenancy was protected by the Act.

Trans-Brittania Properties Ltd v *Darby Properties Ltd* [1986] 1 EGLR 151 CA: The tenant of a garage block sublet the individual garages. Held: Even though he carried out management and dealt with repairs, the tenant was not in occupation of the block and was unprotected.

Graysim Holdings Ltd v *P&O Property Holdings Ltd* [1996] 3 EG 124, HL: The tenant carried on business as market promoters and managers. It held a lease of a market hall, fitted it out as stalls, and granted "licences" of the individual stalls to stall-holders who in fact enjoyed exclusive possession of their stalls. The tenant employed a market superintendent who had a small office in the hall; he controlled access to the hall, although the stall-holders locked up their own stalls. Held: The stallholders were sub-tenants having exclusive possession of their stalls and therefore the tenant could not be treated as being in occupation of the stalls. The tenant's retention of the circulation areas and the ancillary office did not amount to occupation of a holding. The tenant did not have the protection of the Act.

Bassari v *Camden London Borough Council* [1998] EGCS 27, CA: The tenant used the upper floors of a building for letting to residential occupiers. The tenant claimed that he had retained one room for administration of the building but no evidence was provided. Held : Following *Graysim*, the tenant could not be treated as occupying the parts he had sublet. The decision in *Lee-Verhulst* (above) could be distinguished because here the tenant had not retained the keys and a right to enter the rooms. At the most, if there had been evidence in support, he might have had a holding comprising the administration room but that room alone could not be used to continue the business.

2.13 Business purposes

For the protection of the Act to apply, it is necessary for the occupation of the premises, or part of them, to be wholly or partly for the purpose of a business carried on by the tenant.

Where the controlled company and group company provisions of section 23(1A) and section 42 apply (see 2.7 and 2.8), the necessary business use may be that carried on by the person or company whose occupation counts as that of the tenant under those provisions (section 23(1B), section 42(2)). Where the trustee and beneficiary provisions of section 41 apply (see above), the necessary business may be that carried on by the beneficiary whose occupation counts as that of the trustee tenant under those provisions (section 41(1)(a)).

Equivalent provisions are not needed to meet the partnership situation where the tenancy is held in the name of one or more partners (the "business tenants" — see 2.9), since a partnership is not a separate legal entity from its partners and therefore the business of the partnership, carried on in the premises, is in fact the business of the "business tenants", even if they share the ownership of that business with other partners.

The word "business" is defined in section 23(2). It "includes a trade, profession or employment and includes any activity carried on by a body of persons, whether corporate or incorporate". This is a very wide definition; profit-making or the distribution of profits to shareholders are not a necessary feature of "business" although it is a relevant matter when viewing the overall question of whether a "business" is being conducted in the premises (*Hawkesbrook Leisure Ltd* v *Reece-Jones Partnership* [2004] 2 EGLR 61 ChD; *Abernethie* v *AM & J Kleiman Ltd* (1970) 211 EG 405, CA).

The premises themselves need not have the tenant's business actively traded in them. Protection may apply if the occupation is for the purpose of the tenant's business even if that business is primarily conducted elsewhere (*Hillil Property & Investment Co Ltd* v *Naraine Pharmacy Ltd* (1979) 2 EGLR 65). However, the premises must be occupied in furtherance of the business and not just for convenience.

The fact that premises are residential in character, or even used residentially, does not necessarily mean that they are not being occupied for the purpose of a business. Some residents use their accommodation both for living and for a business. In some blocks of flats, a flat is rented from the landlord by the block's management company as accommodation for a caretaker employed to provide those services which constitute the management company's main activity. Whether the activity is a "business" (see 2.3), and whether that business use is sufficiently significant to fall within the Act, are questions of fact and degree;

the amount of business use may be insignificant, in which case it will be ignored. In cases of dispute, this question is a matter for the judge at first instance (*Wright* v *Mortimer* [1996] EGCS 51, CA). There is a special rule where the terms of the tenancy prohibit business use (see 2.14).

The question whether Government activities are a business is discussed at 2.21.

Business can sometimes include the business of granting sublettings of parts of the property itself, even for residential use, but subject to the occupation test mentioned above.

The business use must, for protection to apply, be a significant purpose of the occupation. If it ceases to be, the protection of the Act will cease.

Casenotes

Hills (Patents) Ltd v *University College Hospital Board of Governors* [1955] 3 WLR 523, CA: The Board of Governors of UCH opposed the tenant's application for a new tenancy on the basis of section 30(1)(g) (see Chapter 7), which required an intention that the landlord would occupy the premises for its own business. The tenant claimed that running a NHS hospital was not a "business". Held: It was a business for the purposes of the Act.

Addiscombe Garden Estates v *Crabbe* [1958] 1 QB 513: The premises were let to the tenant for use as a tennis club. The landlord denied that the use was a business. Held: Use as a tennis club was a "business", so the tenant was protected by the Act.

Abernethie v *AM&J Kleiman Ltd* (1970) 211 EG 405, CA An amateur teacher used the loft of his rented home for a weekly Sunday school. He paid over the pupils' subscriptions to a scripture mission. Held: This was not use for a "business" within the Act.

Parkes v *Westminster Roman Catholic Diocese Trustee* [1978] 2 EGLR 50, CA: The church trustee was the landlord of a bowls and tennis club and opposed a renewal of the lease on ground (g) of section 30, namely that the church intended to use the premises for meetings after church services, for religious instruction and for providing recreational facilities to youth and the elderly on a charitable basis. Held: Those proposed uses by the church would amount to a "business" for the purposes of the Act.

Hillil Property & Investment Co Ltd v *Naraine Pharmacy Ltd* [1979] 2 EGLR 65: A store was let to the tenant, who used it in conjunction with his nearby shop. Held: The store was used for the purpose of the tenant's nearby retail business so the Act applied.

Cheryl Investments Ltd v *Saldanha* [1978] 2 EGLR 54: The tenant of a residential flat used it to conduct his importing business, with typewriter, filing and visits by business contacts. Held: He occupied the flat for the purpose of his business and the Act applied.

Royal Life Saving Society v *Page* [1978] 2 EGLR 54: A residential maisonette was let to a medical practitioner who lived there and only rarely saw patients there. Held: He did not occupy it for the purpose of a his business.

Chapman v *Freeman* [1978] 2 EGLR 48: A cottage was let to a nearby hotelier who used it to house some of the staff working at his hotel. Held: (*Obiter*) The tenancy would have been protected by the Act if the tenant used the house as accommodation for staff who were required to live there for the better performance of their duties as employees.

Groveside Properties Ltd v *Westminster Medical School* (1984) 47 P&CR 507: The medical school rented a flat and granted four students licence to live there. The school secretary regularly visited the flat and ensured a mix of students at different stages of their courses, for their educational benefit. Held: The school occupied the flat for the purpose of the school's business.

Gurton v *Parrott* [1990] EGCS 159: A house and outbuildings were let to a tenant who originally used the outbuildings for dog breeding and kennelling. By the time the landlord served a section 25 notice to terminate the tenancy, the tenant's involvement in breeding and kennelling had diminished. Held: The diminution of the business meant that the tenant was no longer in occupation for the purpose of a business, but was in occupation as her residence, so the protection of the Act had been lost.

Wandsworth London Borough Council v *Singh* [1991] 2 EGLR 75: The premises were let to the council for use as a public open space. The council claimed that the Act applied but the landlord argued that the use was not a business. Held: Use by the council for one of its authorised functions was a business, so the council's tenancy was protected by the Act.

Methodist Secondary Schools Trust Deed Trustees v *O'Leary* [1993] 1 EGLR 105: The tenants of a house were the trustees of a nearby school which was run by separate governors. The governors appointed a school caretaker and required him to live in the house, which was connected to the school's alarm system. The tenants claimed that they were occupying the house for the purpose of the school's business. Held: On these facts there was no protection, but if the school had been run by the tenants themselves (rather than by separate governors) and they had engaged the caretaker, then his occupation of the house would have been treated as occupation by the tenants for the purpose of the school business.

2.14 Tenancies which prohibit business use

Section 23(4) of the Act provides that the protection given by the Act will not apply if:

(a) the terms of the tenancy include a prohibition (however worded) against use of the premises for business purposes generally (that is, not just prohibiting a particular business) and
(b) neither the present immediate landlord nor any predecessor of his has given consent to the tenant's use of the premises for his business and
(c) the present immediate landlord has not become aware of that business use and acquiesced in it (that is, took no steps to stop it).

An example would be where a flat is let to a company on a tenancy containing a tenant's covenant not to use the flat for the purpose of any business. If the present landlord knows that the flat is used actively for the purposes of the company's business and he "turns a blind eye", or if a previous landlord actually gave permission for the flat to be used in that way, the protection of the Act may apply notwithstanding the prohibition against business use contained in the tenancy agreement.

The effect of these rules is that a new landlord is not bound by the previous landlord's mere acquiescence in the tenant's use of the premises, but he would be bound by any consent the previous landlord may have given for the business use. The giving of consent requires some positive act by the landlord (*Bell* v *Alfred Franks & Bartlett* [1980] 1 EGLR 56, CA). Sometimes it may be difficult to distinguish between "consent" and mere "acquiescence".

Casenote

Bell v *Alfred Franks & Bartlett* [1980] 1 EGLR 56, CA: The tenancy of a residential flat and garage restricted the use of the garage to the standing of private cars. The tenant actually used the garage to store cartons of samples for its business. The original landlord knew but did nothing to stop it. He sold his reversion and the new landlord gave the tenant a notice to quit but no notice under the Act. The tenant claimed the protection of the Act. Held: "Consent" requires a positive demonstrative act of approval. On the facts, the previous landlord did not consent. Since the present landlord had not acquiesced in the business use, he was entitled to invoke the prohibition against business use in the tenancy and the tenant was not protected by the Act.

2.15 Houses and flats

Where the property let by the tenancy comprises a flat or a house, and it is used for the purpose of a business, the protection of the Act is not excluded merely because the property was constructed for residential use. Even if the terms of the tenancy restrict the use of the property to non-business purposes, the protection under the Act may nevertheless apply where the use for business was the subject of consent by a previous landlord or the subject of consent or acquiescence by the current landlord, as discussed at 2.14. If protection exists under those rules, or if the terms of the tenancy actually permit business use, then the protection of the Act will apply but subject to some additional exceptions and with some modification of the notice procedures.

There are three sets of exceptions to the application of the Act in relation to these types of property. First, some premises comprised in tenancies are excluded from the protection of the 1954 Act by section 24(2) of the Rent Act 1977. While the provisions of section 24(2) appear somewhat complex, they effectively exclude from the protection of the 1954 Act any dwelling let on a protected or statutory tenancy under the Rent Acts which falls within the category of a "controlled tenancy" or which would have fallen into that category were it not a tenancy at a low rent. Further detail on this topic would be beyond the scope of this book. Second, the Act will not apply to the tenancy of a house if the tenancy comprises an extended lease which was granted under section 14 of the Leasehold Reform Act 1967 (1967 Act section l6(1(c)). Third, the Act will, similarly, not apply to the tenancy of a flat if the tenancy comprises a new lease which was granted under Chapter II of Part I of the Leasehold Reform, Housing and Urban Development Act 1993 (1993 Act, section 59(2)). In each of these cases, the exclusion also prevents a subtenant from having the protection of the 1954 Act when such a tenancy ends.

Where a tenancy of a house is within the protection of the 1954 Act, there are special notice procedures, discussed at 6.10.

2.16 Contracting out

Section 38(1) of the Act provides that even if the landlord and the tenant agree, as a term of the tenancy or separately, that the protection of the Act will not apply, this will generally be ineffective and will not in fact deprive the tenant of protection. However, section 38A(1) and (3) provide a procedure (commonly referred to as "contracting out") which, if strictly followed, allows the tenancy to contain such an agreement and for it to be effective. This is explained in detail in Chapter 3.

The decision to contract out should not be taken lightly. Security of tenure protects the goodwill of the business and this will be lost if the landlord refused to renew the lease, or may be severely damaged if an excessive rent is demanded for a voluntary renewal. Where goodwill is valuable to the tenant, the landlord can effectively blackmail the tenant as to the new lease terms if there is no statutory right of renewal. Contracting out should only be contemplated by a tenant where a short term tenancy is acceptable or where moving premises is easy and cheap.

From the landlord's point of view, contracting out a long letting would be detrimental at rent review. Contracting out can only realistically be justified for a landlord in the case of short term lettings or where possession will be required at some future date and the landlord does not want to face arguments about, for example, the validity of his intention to redevelop (see 8.7). Contracting out would be particularly helpful where the landlord may want possession in order to carry out a refurbishment scheme which falls short of substantial redevelopment.

2.17 Short fixed-term lettings

Section 43(3) of the Act provides that the protection of the Act will not apply to a tenancy granted for a term certain of up to six months unless:

(a) the tenancy contains provisions for renewing the term or extending it beyond six months from its beginning or
(b) the tenant has been in occupation for a period which, together with any period during which any predecessor in the carrying on of the same business was in occupation, exceeds 12 months.

The effect of this is that a new tenant can be given a fixed-term tenancy for six months and, when it expires, a further fixed-term tenancy for just under six months, without attracting the protection of the Act, provided that the terms of the initial fixed-term tenancy did not include a legal right for the tenant to be granted the second fixed-term tenancy. If the tenant is allowed to remain in occupation after the end of the second short term tenancy and this takes his period of occupation to over 12 months, he will acquire protection if his continued occupation is by virtue of a tenancy. If however he is continuing in occupation either as a trespasser against the landlord's wishes or merely as a tenant at will (for example, while negotiations are taking place for him to be granted a proper contracted-out tenancy — see 2.3) that does not count as occupation for section 43(3) and he will not acquire protection (*Cricket Ltd* v *Shaftesbury plc* [1999] 2 EGLR 57, ChD).

Some caution is needed when applying this rule. The exclusion of protection only applies to fixed-term lettings for up to six months; a periodic (eg monthly) letting will be fully protected even if notice to end it is given within the first six months.

It should be possible to include a break option in a short tenancy of this type — see the discussion about "term certain" at 3.1.

2.18 Agricultural land and mines

The protection of the Act is not given to a tenancy of an agricultural holding which is protected by the Agricultural Holdings Act 1986 (1954 Act section 43(1)(a)). The 1986 Act defines an agricultural holding as:

> the aggregate of the land (whether agricultural land or not) comprised in a contract of tenancy which is a contract for an agricultural tenancy,

but it excludes land let to a tenant during the continuance of his office, appointment or employment with the landlord (1986 Act, section 1(1)).

Protection under the 1954 Act is also withheld from a tenancy which would be an agricultural holding under the 1986 Act but is excluded from the protection of the 1986 Act by section 2(3) of that Act or by an approval was given under section 2(1) of that Act. This means that the 1954 Act does not apply to agricultural tenancies even if they are outside the 1986 Act, provided that the reason for their exclusion from the 1986 Act is simply because they grant an interest less than a tenancy from year to year, or grant a fixed term of more than one year but less than two years (see *EWP Ltd* v *Moore* [1991] 2 EGLR 4, CA), or are limited to seasonal grazing or mowing, or have been approved by the Minister for exclusion from the protection of the 1986 Act.

A farm business tenancy under the Agricultural Tenancies Act 1995 is also excluded from the 1954

Act. In brief, a farm business tenancy is a tenancy of land granted on or after 1 September 1995 and continuously farmed for the purpose of a trade or business, where either the character of the tenancy is wholly or primarily agricultural or else that was its character at the beginning of the tenancy and the parties agreed before its commencement that it would be treated as a farm business tenancy throughout its continuance.

Difficulties sometimes occur in distinguishing between agricultural use and business uses such as garden centres and horse stabling and rearing. Use for horse grazing is an agricultural use falling within the 1986 Act, whereas use for horse training or as a riding school is a non-agricultural business which might be protected by the 1954 Act (*Rutherford* v *Maurer* [1962] 1 QB 16; *Bracey* v *Read* [1963] Ch 88; *Watkins* v *Emslie* [1982] 1 EGLR 81, CA). Where the use of the land has changed from agricultural use, the court may be reluctant to find that the tenancy has become a business tenancy within the 1954 Act without clear evidence that the agricultural use has been abandoned in favour of a non-agricultural business use (*Wetherall* v *Smith* [1980] 2 EGLR 6, CA). Where the use is mixed, the most substantial land use will usually dictate the applicable statutory regime (*Lord Monson* v *Bound* (1954) 164 EG 377; *Short* v *Greeves* [1988] 1 EGLR 1).

Mines let on a mining lease are excluded from the protection of the Act by section 43(1)(b). Mining lease is defined in section 25 of the Landlord and Tenant Act 1927 as:

> a lease for any mining purpose or purposes connected therewith and 'mining purpose' includes the sinking and searching for, winning, working, getting, making merchantable, smelting or otherwise converting or working for the purposes of any manufacture, carrying away, and disposing of mines and minerals, in or under land, and the erection of buildings, and the execution of engineering, and other works suitable for those purposes.

In addition to coal and similar mining, this includes the mining of all substances capable of being worked for profit below the surface of the land, such as gravel, sand and clay, but not peat (*O'Callaghan* v *Elliott* [1966] 1 QB 601).

Casenotes

Rutherford v *Maurer* [1962] 1 QB 16: Land was let for the grazing of horses belonging to a riding school. Held: This was an agricultural use, so the tenancy was not protected by the Act.

Bracey v *Read* [1963] Ch 88: Land was let for the training and exercising of racehorses. Held: This was a business use, not an agricultural use, and so the tenancy was protected by the Act.

Short v *Greeves* [1988] 1 EGLR 1: 6.2 acres of land and a garden centre were let on a yearly tenancy within the Agricultural Holdings Act 1948. Over the years, the volume of sales of bought-in produce, as compared with produce grown on the premises, increased to 60%. The landlord served a termination notice under section 25 of the Act but the tenant claimed that his tenancy was still an agricultural tenancy. Held: It was an agricultural tenancy. The business was still in large part based on the home-grown produce. It was not within the 1954 Act.

2.19 Service and employment tenancies

Section 43(2) provides that the protection of the Act is not given to a tenant who is given his tenancy by reason of his being the holder of an office, appointment or employment, if it is a term of the tenancy that it will only continue so long as the tenant holds that office etc., or it is terminable by the landlord on the tenant ceasing to hold the office, or it comes to an end at a time fixed by reference to the time at which the tenant ceases to hold that office.

In relation to such a tenancy which was granted after the 1954 Act originally came into force on 1 October 1954, the Act is excluded only if the tenancy was granted in writing and sets out the purpose for which it was granted.

Care must be taken with such tenancies. If there is any doubt as to the linkage between the tenancy and the tenant's office or employment, it would be better to use a contracted out tenancy.

2.20 Licensed premises

Originally the protection of the Act was not given where the tenancy comprised premises licensed for the sale of intoxicating liquor for consumption on the premises. That exclusion was abolished by the Landlord and Tenant (Licensed Premises) Act 1990. Since 11 July 1992 the exclusion has ceased to apply to any tenancy, whenever it may have been granted.

2.21 Government as tenant

The position of Government bodies as tenants of business premises is subject to special provisions of the Act, because generally Acts of Parliament do not bind "the Crown" (which includes Her Majesty's Government) except so far as expressly provided, and because of the question which would arise as to whether Government activities are a "business".

Section 56(3) provides that Part II of the Act will apply where a tenancy is held by or on behalf of a Government department if the whole or part of the premises are occupied for any purposes of a Government department. In such cases the premises are deemed to be occupied for the purpose of the tenant's business. This also applies if the Government department which is the tenant provides the premises rent-free to any other person (section 56(4)).

Further, section 56(3) also provides that, on any change of Government occupier, there is a deemed succession to the business if the test of occupation for the purposes of a Government department is met both before and after the change of occupier. Thus a change from, say, a job centre to a driving licence centre will be treated as continued occupation for the same business. This will be relevant, for example, to the five year and 14 year tests in relation to statutory compensation (see 14.2 and 14.3).

In *Linden* v *Department of Health and Social Security* [1986] 1 EGLR 108, ChD, it was held that a tenancy to the Secretary of State was protected by the Act where the premises were used as furnished flats let to NHS hospital staff under the close and active management of the district health authority, which was administering the functions of the Secretary of State.

2.22 Other exclusions

Some other statutes contain provisions that the 1954 Act will not apply to particular tenancies referred to in those statutes. In outline, these exclusions include:

(a) tenancies to dockyard contractors of land in a Royal dockyard, if designated as excluded from the 1954 Act by the Secretary of State under the Dockyard Services Act 1986
(b) concession leases granted by the Secretary of State to persons constructing or operating the tunnel system under the Channel Tunnel Act 1987

(c) tenancies granted to contractors of land at premises designated under the Atomic Weapons Establishment Act 1991
(d) tenancies of land comprised in a railway network or facility arising under an agreement relating to the grant of a franchise to provide services using the network or facility under the Railways Act 1993
(e) tenancies granted under section 30 of the Armed Forces Act 1996 in respect of the Royal Naval College, the Dreadnought Seaman's Hospital or the Devonport Nurses Home.

2.23 Crown property

An Act of Parliament will not generally bind the Crown (including government departments) unless it contains express provisions making it apply. The 1954 Act contains provisions making it binding on the Crown, both as landlord and as tenant. The provisions which give government departments the protection of the Act as tenant have been considered at 2.21.

In the case of premises where the landlord is the Crown or a government department, this fact does not prevent a tenant from having the protection of the Act (section 56(1)). For this purpose, the Eighth Schedule to the Act makes provision for Her Majesty to be represented, as landlord, by the Chancellor of the Duchy of Lancaster in relation to property held in right of that Duchy and for the Duke of Cornwall to be treated as landlord in respect of property belonging to the Duchy of Cornwall. However, government departments and other public bodies who are landlords have, in specified circumstances, special rights to terminate tenancies, to oppose renewals or to require special provisions to be inserted into new tenancies under the Act (see 6.9 and 12.5).

Contracting Out

3.1 Leases which may be contracted out

The current contracting out procedure and the old procedure, both described below, apply only where the tenancy to be granted will be "for a term of years certain" (new section 38A(1), old section 38(4)(a)). A purported contracting out of a tenancy which is not "for a term of years certain" will be ineffective and the tenancy will be protected.

The phrase "term of years" can include a fixed term of less than a year (*Re Land and Premises at Liss, Hants* [1971] Ch 986).

The word "certain" excludes periodic tenancies and also excludes a tenancy which has an initial fixed term and then is stated to continue until terminated by notice (*Nicholls* v *Kinsey* [1994] 1 EGLR 131, CA). On the other hand, a fixed term which can be ended earlier under a break option exercisable by notice to be given by a party will be treated as a "term of years certain" for this purpose; the existence of the break clause is immaterial. This is because section 69 defines "notice to quit" as including "a notice to terminate ... a tenancy for a term of years certain", so the Act contemplates that a term of years can be "certain" even if it is terminable by a notice to quit (see *Scholl Manufacturing Co Ltd* v *Clifton (Slim-Line) Ltd* [1967] Ch 41, CA *per* Diplock LJ; *Receiver for the Metropolitan Police District* v *Palacegate Properties Ltd* [2000] 1 EGLR 63, CA).

3.2 Agreements excluding sections 24 to 28

The agreement permitted by the contracting out procedure is limited to an agreement which excludes sections 24 to 28 from applying to the tenancy, rather than one which excludes the whole of Part II of the Act from applying. Nevertheless this will have the effect of excluding all the protection given by the Act, since sections 24 to 28 cover both the tenancy continuation rules and the termination notice rules; furthermore, the compensation provisions of the Act operate only where the termination notice rules apply, so they too will be excluded by such an agreement.

3.3 The two procedures for contracting out

The procedure for contracting out has been changed by the 2004 amendments. The current procedure is described at 3.4.

Some leases contain a tenant's covenant not to sublet the premises, or perhaps part only of the premises, unless the subtenancy excludes to protection of the Act. As from 1 June 2004, the transitional provisions of the RRO require any such provisions, where they refer to the old contracting out procedure, to be read as if they referred to the new contracting-out procedure (RRO article 29). This applies whether the lease containing the references to the old procedure was granted before or after 1 June 2004.

However, the old procedure is still available in certain circumstances, described at 3.5 below. A possible lacuna in the transitional provisions is also discussed below.

3.4 Current procedure for contracting out

The procedure for contracting out under section 38A(3) is now set out in the Schedule 2 of the RRO. There are however a number of unanswered questions relating to this new procedure. The procedure involves taking three steps.

The first step is for the landlord to give the tenant a warning notice in the form set out in schedule 1 of the RRO (section 38A(3)(a)). This form of notice has blank spaces to be completed by the landlord with the names and addresses of the landlord and the tenant. There is no space on the form for inserting any details of the property to be let or any terms of the letting; the purpose of the notice is simply to warn the tenant of the consequences of contracting out and to advise the tenant to seek professional advice.

Some take the view that the notice should be accompanied by details of the property and a draft lease, but there is nothing in the Act or the RRO to suggest that this is either necessary or desirable. Indeed, doing so might lead to other complications; for example, if a draft lease is attached to the notice and the draft is then amended, it might be necessary to serve a fresh notice. Further, attaching anything to a prescribed form of notice might arguably invalidate the notice. The authors' view is that the prescribed notice should not have anything attached to it or given with it apart from, where appropriate, a suitable covering letter identifying the premises to be let.

The second step is for the tenant to make a written declaration, referring to the proposal to enter into the lease and acknowledging that he has received the warning notice (section 38A(3)(b), invoking schedule 2 of the RRO). There are two alternative prescribed forms to be used for the declaration. The form set out in para 7 of schedule 2 of the RRO is to be used if there is at least a 14 day gap between the date when the landlord gave the warning notice to the tenant and the date on which the lease is completed or, if this is preceded by an agreement for lease, the date on which the agreement is exchanged. That form can simply be signed by the tenant. If the gap is less than 14 days, the tenant must instead make a statutory declaration in the form set out in para 8 of schedule 2 to the RRO, and this declaration must be administered and witnessed by an independent solicitor or commissioner for oaths. It is understood that the purpose of requiring a statutory declaration to be made where there is no 14 day cooling-off period is to bring home to the tenant the significance of contracting out. The question arises whether, having arranged for a statutory declaration to be made in accordance with para 8 at a time when the lease was expected to be completed in less than 14 days from the giving of the warning notice, that statutory declaration can still be relied upon if the completion of the lease is delayed beyond the end of the 14 day period, or whether it is then necessary to obtain a further, ordinary, declaration from the tenant under para 7. Commonsense would dictate that a fresh declaration should not be needed but the RRO appears to require it. It is not clear whether one could, for this purpose, rely on the encouragement given by the Court of Appeal not to take an "over-technical" approach to the Act (*Brighton and Hove City Council* v *Collinson* [2004] 2 EGLR 65, CA, relating to the old contracting-out procedure).

Both the forms of declaration require insertion of the names of the landlord and the tenant, the address of the premises to be let, and the date on which the term is to commence. The blank space provided for the insertion of the commencement date does not have a year number at the end of the space and it appears therefore that it is not necessary to insert a specified calendar date — the commencement date could be expressed by reference to some agreed provision (for example, "the date to be determined under clause X of the agreement for lease") or perhaps "a date to be agreed".

Both the declaration forms also include a reproduction of the blank form of prescribed warning notice — not a reproduction of the actual notice that the landlord gave, so the blanks spaces are not to be filled in.

The declaration forms make provision for the person making the declaration to be either the tenant or someone duly authorised by the tenant to make the declaration. That authority should be given in writing. Where the tenant is a company, ideally the directors should resolve to give someone that authority, although it may be the case that a director or other officer of the company has implied authority to deal with this sort of document as part of the day to day business of the company. The RRO does not actually require the declaration or the evidence of the declarant's authority to be handed over to the landlord, but it is normally the landlord who is most concerned to see that the contracting out is valid and therefore the landlord should insist that the declaration and a copy of any written authority in favour of the declarant are given to him and he should keep them with the counterpart lease following the grant of the lease.

The third step is to complete the lease or exchange an agreement for lease, inserting into the lease or indorsing on it both the agreement that sections 24 to 28 are excluded and a reference to the warning notice and the tenant's declaration (RRO schedule 2 paras 5 and 6).

The following wording may be suitable for these clauses :

The Landlord and the Tenant —

(a) agree that the Landlord served on the Tenant on [20] a notice as required by section 38(A)(3(a) of the Landlord and Tenant Act 1954 applying to the tenancy created by this lease [and a copy of that notice is annexed to this lease]; and

(b) agree that [the Tenant] [name of declarant] [who was duly authorised by the Tenant to do so] made a [statutory] declaration dated [20] in accordance with the requirements of Section 38(A)(3)(b) of the said Act [and a copy of that declaration is annexed to this lease] and

(c) agree that the provisions of Section 24 to 28 of the said Act are excluded in relation to the tenancy created by this lease.

The warning notice must be given and the appropriate declaration must be made before the tenant enters into the tenancy or becomes contractually bound to do so (RRO schedule 2 paras 2 and 4). The reference to the tenant becoming "contractually bound" to enter into the tenancy presumably means, primarily, entering into an agreement for lease. Accordingly, the procedure for contracting out must take place before the lease is entered into and, if that is preceded by an agreement for lease, before the agreement for lease is exchanged. One cannot, therefore, have an agreement for lease which validly imposes an obligation on the tenant to comply with the contracting out procedure — the procedure must be carried out before exchanging the agreement.

If there is some change in the arrangements between the parties during the period between the making of the declaration and the completion of the lease, it may possibly be necessary or desirable for a fresh declaration to be made. A fresh declaration is most likely to be desirable if the identity of the landlord or the tenant has changed from the persons named in the declaration. This might occur, for example, where the landlord's interest is transferred to a new landlord or where the tenant decides to take the lease in the name of another company in its group. In such cases the statement in the declaration about the intention of the named parties to enter into the tenancy will patently have been superseded and a fresh declaration may be needed. A problem would arise for the landlord if the tenant (or the proposed new tenant) refused to make the fresh declaration and the landlord was nevertheless bound to grant the lease, since that lease might not be validly excluded from protection. Accordingly, if a landlord is to enter into an agreement for a contracted-out lease relying on a declaration made under the new procedure, it may be necessary to provide in the agreement that the tenant may neither assign the benefit of the agreement nor require the lease to be granted to another person. Such provisions may be perfectly acceptable in many cases but not invariably. An alternative might be to provide in the agreement for lease that the landlord cannot be required to grant the lease to a different tenant unless that tenant invites the landlord to give him a warning notice and, if the landlord gives it to him, he makes the appropriate declaration. Provided that the wording does not compel the new intended tenant to follow that procedure, but merely makes it a precondition of being able to demand that the lease is granted to him instead of to the originally intended tenant, such a provision may be enforceable.

If some other terms of the proposed tenancy change, or had not been finalised when the declaration was made, the desirability of obtaining a fresh declaration is less clear, since the prescribed wording of the declaration does not mention the intended tenancy being in any particular form.

There is a related issue where a contracted-out lease is to contain a covenant by a guarantor and, as is quite common, the covenant is to require the guarantor to take up, if required by the landlord, a substitute lease of the premises in the event that the initial lease is disclaimed upon the tenant becoming insolvent. The landlord, having taken the trouble to ensure that the lease to the tenant is excluded from the protection of the Act, will probably want any substituted lease to be similarly contracted out. However, inserting a provision requiring the guarantor to join with the landlord in following the new contracting-out procedure before taking up the substitute lease will not deal satisfactorily with the matter because such a provision will by itself almost certainly be unenforceable — and indeed will be too late — for the reason given above. Consequently in that eventuality the landlord might have to choose between giving the guarantor a substitute lease which is protected by the Act or not invoking the guarantor's obligation to take a substitute lease. This decision should be made using valuation criteria. If the purpose of contracting out of the current lease had been to ensure that the landlord could easily obtain possession for, say, refurbishment after a relatively short period has elapsed, contracting out may be valuable and important to the landlord if those requirements continue to apply. However, if the unexpired term is relatively long and there was no compelling need to be able to regain possession easily on its expiry, it might be worthwhile requiring the guarantor to take a new lease even if it is inside the Act.

In order to ensure that such a substitute lease to a guarantor is effectively excluded from protection, it has been suggested that, in addition to giving the statutory warning notice to the intended tenant, such a notice should also be given to the intended guarantor and that both the tenant and the guarantor should make the appropriate declarations. The guarantor's declaration should refer to the lease to be granted pursuant to the proposed guarantee covenant and will therefore be a lease which will commence on "the date to be determined under clause X of the lease to be granted to [the tenant]", being a reference to the guarantee clause. However, this might not be an effective solution in all circumstances; for example, the declaration has to name the intended landlord and therefore would not remain correct if the identity of the landlord changes at a later date but before the time when the guarantor's covenant to take the substitute lease has to be invoked.

Another issue is the position where the landlord of contracted-out premises is required under section 19 of the Landlord and Tenant (Covenants) Act 1995 to grant an overriding lease of the premises to a former tenant from whom the landlord has obtained payment of the current tenant's arrears of rent. While the clear intent of the 1995 Act is to make the overriding lease correspond with the current lease, neither the 1995 Act nor the 1954 Act make express provision for an overriding lease to be contracted out in these circumstances, so the position is not wholly certain.

3.5 Old court application procedure

Before the 2004 amendments, the contracting out procedure (under what was section 38(4)(a)) required the parties jointly to obtain a court order authorising them to enter into the exclusion agreement. The court order had to be obtained before the lease was completed in order to validate the exclusion agreement (*Essexcrest Ltd* v *Evenlex Ltd* [1988] 1 EGLR 69). The transitional provisions of article 29 of the RRO provide that any such court orders obtained before 1 June 2004 (when the RRO amendments came into force) remain effective after that date and also provide that the court order procedure remains available for use by parties who have to obtain a court order in order to comply with an agreement for lease entered into before 1 June 2004. Such agreements, under the old law, were frequently conditional upon obtaining the court order between exchange and completion.

Unfortunately there appears to be an omission from these transitional provisions. They do not expressly cater for the situation where a person has an option (as distinct from an agreement) to take up a new contracted-out lease and the provisions of the option require the parties to obtain a court order under (old) section 38(4)(a). An example of this might be where a tenant has an existing contracted-out short lease containing an option for him to take another short contracted-out lease when it expires. On the traditional legal view of options for leases, an "agreement" for the new lease will only come into existence when the option is exercised. If that happens after 1 June 2004, it will be outside the transitional provisions of the RRO and the court order procedure will not be available. Whether the result would be that the tenant cannot insist on being granted the new lease, or the landlord cannot insist on excluding it from the protection of the Act, may depend on the wording of the option. This is not covered by the separate transitional provision of the RRO by which references in leases to the old court procedure for contracting out are now to be read as references to the new procedure, since that is restricted to provisions relating to subletting (RRO article 29(3)).

The joint application to the court is a simple procedure in which the solicitors to each party sign and submit to the court a joint application for an exclusion order together with copies of the agreed draft lease, an agreed form of court order and the court fee. Usually this is done through the post without personal appearance in court. Since the court application has to exhibit the draft lease and the court order will refer to the draft, at least in relation to the contracting out clause, the question has arisen whether changing the text of the lease (in relation to matters other than contracting out) between applying for the court order and completing the lease invalidates the contracting out unless a fresh court order is obtained referring to the amended draft lease. It has been held that changes which are not material to the contracting out, even in certain circumstances a change in the parties to the lease, will not necessarily invalidate the contracting out; furthermore, the Court of Appeal has commented that the Act should not be treated over-technically (*Receiver for the Metropolitan Police District* v *Palacegate Properties Ltd* [2000] 1 EGLR 63, CA; *Brighton and Hove City Council* v *Collinson* [2004] 1 EGLR 65, CA).

In one case under the old procedure, an existing landlord and tenant entered into an agreement for a contracted-out lease which was conditional on obtaining the court order. They then obtained the order, thus fulfilling the condition, and the tenant continued in occupation. However, the parties inadvertently omitted to complete the new lease. The court held that the tenant's current tenancy nevertheless did contain the exclusion of protection that had been sanctioned by the court order, under the ancient legal doctrine of *Walsh* v *Lonsdale* (1882) 21 CLD 9 by which the agreement for lease was treated as granting the lease in the agreed form (*Tottenham Hotspur Football & Athletic Co Ltd* v *Princegrove Publishers Ltd* [1974] 2 QB 17).

3.6 Assignment of contracted-out leases

A lease which contains a valid exclusion of sections 24 to 28 of the Act retains the effectiveness of that exclusion when it is assigned to a new tenant.

This applies not only to an express written assignment but also to an assignment by operation of law, such as where the tenant purports to grant a sublease of the premises for the whole residue of the term of his lease without reserving any reversion (*Parc Battersea Ltd* v *Hutchinson* [1999] 2 EGLR 33, ChD).

3.7 Authorising agreements for surrender

Where a tenancy already exists which is protected by the Act and the parties wish to enter into an agreement for the tenancy to be surrendered at a later time, the effect of section 38 of the Act also needs considering. As will be seen from 4.3 below, such an agreement for surrender will generally be void if, at the time of making the agreement, the tenancy falls within the protection of the Act (section 38(1); *Joseph* v *Joseph* [1966] 3 WLR 631).

However, section 38A(2) and (4) provides a procedure (similar to the contracting-out procedure mentioned above) which, if strictly followed, allows the tenant to enter into an effective agreement for surrender. The procedure for this contracting out is set out in schedule 4 of the RRO. It involves taking three steps.

The first step is for the landlord to give the tenant a warning notice in the form set out in schedule 3 of the RRO (section 38A(4)(a)). This form of notice has blank spaces to be completed by the landlord with the names and addresses of the landlord and the tenant. There is no space on the form for inserting any details of the property to be surrendered; the purpose of the notice is simply to warn the tenant of the consequences of entering into an agreement for surrender and to advise the tenant to seek professional advice. Some take the view that the notice should be accompanied by details of the property and a draft agreement for surrender, but there is nothing in the Act or the RRO to suggest that this is either necessary or desirable. Indeed, doing so might lead to other complications; for example, if a draft agreement is attached to the notice and the draft is then amended, it might be necessary to serve a fresh notice. Further, attaching anything to a prescribed form of notice might arguably invalidate the notice. The authors' view is that the prescribed notice should not have anything attached to it or given with it apart from, where appropriate, a suitable covering letter.

The second step is for the tenant to make a written declaration, referring to the proposal to enter into the agreement for surrender and acknowledging that he has received the warning notice (section 38A(4)(b) and schedule 4 of the RRO). There are two alternative prescribed forms to be used for the declaration. The form set out in para 6 of schedule 4 of the RRO is to be used if there is at least a 14 day gap between the date when the landlord gave the warning notice to the tenant and the date on which the agreement for surrender is entered into. That form can simply be signed by the tenant. If the gap is less than 14 days, the tenant must instead make a statutory declaration in the form set out in para 7 of schedule 4 to the RRO, and this declaration must be administered and witnessed by an independent solicitor or commissioner for oaths. It is understood that the purpose of requiring a statutory declaration to be made where there is no 14 day cooling-off period, is to bring home to the tenant the significance of entering into an agreement for surrender.

Both forms of declaration require insertion of the names of the landlord and the tenant and the address of the premises contained in the tenancy to be surrendered. The declaration forms include a reproduction of the blank form of prescribed warning notice (not the actual notice, so the blanks spaces are not to be filled in).

The declaration forms make provision for the person making the declaration to be either the tenant or someone duly authorised by the tenant to make the declaration. Any such authority should be given in writing. Where the tenant is a company, ideally the directors should resolve to give someone that authority, although it may be the case that a director or other officer of the company has implied authority to deal with this sort of document as part of the day to day business of the company. The RRO does not actually require the declaration or the evidence of the declarant's authority to be handed over to the landlord, but it is normally the landlord who is most concerned to see that the agreement for surrender is valid and therefore the landlord should insist that the declaration and a copy of any written authority in favour of the declarant are given to him and he should keep them with the part of the agreement for surrender signed by the tenant.

The third step is to enter into the agreement for surrender, inserting into it or indorsing on it a reference to the warning notice and the tenant's declaration (RRO schedule 4 para 5).

The warning notice must be given and the appropriate declaration must be made before the tenant enters into the agreement for surrender or, if earlier, "becomes contractually bound to do so" (RRO schedule 4 paras 2 and 3). The reference to the tenant becoming "contractually bound" to enter into the agreement for surrender might encompass, for example, giving notice under a surrender-back clause in a lease which requires the tenant to offer a surrender to the landlord before seeking consent to assignment (see 3.8). Accordingly, in order to have a valid agreement for surrender, the procedure under schedule 4 of the RRO must take place before the agreement for surrender is exchanged or any other contractual process leading to it is triggered. This means that one cannot have an agreement for surrender which is conditional on the parties performing the validation procedure — the procedure must be carried out before exchanging the agreement.

Section 38A(2) applies to "the persons who are the landlord and the tenant", so this procedure applies only where the lease already exists and it can be used only by the current parties. Further, it applies only to an agreement to surrender the whole tenancy and not just part of the premises. Hence it cannot be used, say, to validate an agreement for surrender betweeen the tenant and a prospective purchaser of the landlord's interest, nor to validate the insertion into a proposed non-contracted out lease of a clause entitling the landlord to require the tenant to surrender part only of the premises.

Before the 2004 amendments, the validation procedure under the old section 38(4)(b) required the parties jointly to obtain a court order authorising them to enter into the agreement for surrender. That process is no longer available.

A landlord having a tenant who is willing to vacate at the end of his contractual term may want to have the certainty of being entitled to obtain possession from the tenant at that time. This need may arise, for example, where the landlord wants to enter, before the end of that term, into an agreement for lease with a new tenant. In such cases, the landlord should consider simpler alternatives to entering into an agreement for surrender with the present tenant under the RRO procedure. One alternative is to arrange for the tenant to serve a notice on the landlord under section 27(1) of the Act (see 4.6). This can be done at any time before the last three months of the contractual term and will have the effect of taking the present lease outside the protection of the Act.

3.8 Surrender back clauses

Some leases contain provisions that the tenant cannot assign the lease unless he first makes an offer to the landlord to surrender the lease, and he will be entitled to assign the lease only if the landlord fails to notify the tenant within a specified period that he accepts the offer. If the tenant makes the offer and the landlord notifies him that he accepts it, under traditional contract law such an offer and acceptance would create an agreement to surrender the lease.

There are two legal issues which arise here. One is the question whether such an agreement would be valid under section 2 of the Law of Property (Miscellaneous Provisions) Act 1989, which is outside the scope of this book. The other issue, arising under section 38(1) of the 1954 Act, is that an agreement for surrender will be void if it is made at a time when the tenancy is protected by the Act unless the authorisation process under s.38 has been followed (see 3.7 and *Alnatt London Properties* v *Newton* [1983] 1 EGLR 1, decided under the old law). The consequence of section 38, in this situation, would be that neither party would be able to compel the other to complete the surrender which they had purportedly agreed to complete. Whether the tenant would be entitled to assign the lease to a third party if the landlord, having purportedly accepted the offer, declines to complete the surrender will depend on the exact wording of the clause.

The question arises whether, following the 2004 amendments, a surrender back clause could be framed so as to enable or require the parties to observe the RRO procedure prior to the making of the offer and its acceptance. As discussed at 3.7, the provisions of para 2 of schedule 4 require the warning notice to be given, and para 3 requires the tenant's declaration to be made, before the tenant enters into the agreement for surrender or, if earlier, before he "becomes contractually bound to do so".

The cautious view is that the surrender back clause itself makes the tenant "contractually bound" to enter into an agreement for surrender, albeit conditionally upon the tenant desiring to assign the lease and the landlord being willing to accept the offer of surrender. On that view, once the lease containing the surrender back clause has been granted, it will be too late to give the warning notice or obtain the declaration. The warning notice and declaration would need to have been given and made before completing the lease, or exchanging an agreement for that lease, in order that the agreement for surrender arising from the surrender back clause can be said to have been authorised under the RRO. However, that would not be possible since this procedure is only available to persons who are already the landlord and the tenant in relation to the tenancy to be surrendered, so it cannot be used before the lease is granted (section 38A(2)). On that basis, surrender back clauses may still fall foul of the Act.

The less cautious view is that the RRO procedure simply has to be followed before the surrender offer is made or accepted. However, it may be invalid for the lease to purport to impose an obligation on the tenant to go through the RRO procedure. Instead one might suggest using the power under section 19(1A) of the Landlord and Tenant Act 1927 allowing landlords to set out circumstances for refusing consent to assignment. One such circumstance could be the tenant's failure to comply with schedule 4 of the RRO and to make the surrender back offer. Framed in that way, the tenant would not be placed under an actual obligation to comply with the RRO procedure, but his right to seek consent for assignment would simply not arise unless he did in fact follow that procedure.

For example, a provision might be drafted stating that the landlord may refuse consent to an assignment unless:

(a) on or before a date not less than 30 working days before making an application for consent for that assignment, the tenant shall have given the landlord a preliminary notice in writing informing the landlord of his intention to give the landlord an offer-back notice and inviting the landlord to give him a warning notice in the form in schedule 3 of the RRO within the following ten working days and
(b) if the landlord shall have given such warning notice within that period of ten working days (but not otherwise), the tenant shall within the ten working days after the giving of the warning notice have made the appropriate declaration in the form set out in schedule 4 of the RRO and shall have delivered it to the landlord together with a signed offer-back notice in a specified form (drafted to facilitate comply with section 2 of the Law of Property (Miscellaneous Provisions) Act 1989) inviting the landlord to accept it within the following ten working days and
(c) if the above shall have taken place (but not otherwise), the landlord shall have rejected or failed to accept the offer of surrender contained in the offer back notice within those ten working days.

Whether this device will be held to be lawful under section 38 of the Act and schedule 4 of the RRO remains to be determined.

4 Continuation and Termination of Tenancies

4.1 Tenancy continues unless properly terminated

If a tenant has the protection of the Act (see Chapter 2), he will be entitled to remain in occupation for business purposes in all or part of the premises after the end of the term of his lease, and the tenancy created by that lease will be continued by virtue of section 24, unless:

(a) the term of his lease was brought to an end by surrender by the tenant, or by forfeiture of the lease by the landlord, or by forfeiture of a superior lease by a superior landlord, or by a contractual notice to quit or break notice given by the tenant (section 24(2)) or
(b) the tenant gave the landlord a notice under section 27(1) that he did not want the tenancy to continue (see 4.6) or
(c) a section 25 termination notice (whether or not opposing a renewal) was given by the landlord (see Chapter 6) specifying the last day of the term of the lease as the termination date and by that date the tenant failed to reach agreement with the landlord for a new tenancy (see Chapter 13) and neglected to ensure that a court application was made or an extension of time was agreed (see Chapter 10) or
(d) a section 26 request for a new tenancy was given by the tenant (see Chapter 7) specifying the last day of the term of the lease as the date on which his new tenancy is to start, and by that date the tenant failed to reach agreement with the landlord for a new tenancy (see Chapter 13) and neglected to ensure that a court application was made or an extension of time was agreed (see Chapter 10).

Once the tenant is continuing in occupation under a continuation tenancy (commonly called "holding over"), his continuation tenancy can be ended only by:

(a) the continuation tenancy being brought to an end by surrender by the tenant, or by forfeiture by the landlord, or by forfeiture of a superior lease by a superior landlord (section 24(2)) or
(b) by the tenant giving the landlord a notice under section 27(2) to end the continuation tenancy (see 4.7) or
(c) by the landlord giving the tenant a section 25 termination notice, whether or not opposing a renewal (see Chapter 6), and the tenant failing to reach agreement with the landlord for a new

tenancy (see Chapter 13) and neglecting to ensure that a court application is made or an extension of time is agreed (see Chapter 10) by the specified termination date or

(d) by the tenant making a section 26 request to the landlord for a new tenancy (see Chapter 7) and the tenant failing to reach agreement with the landlord for a new tenancy (see Chapter 13) and neglecting to ensure that a court application is made or an extension of time is agreed (see Chapter 10) by the day before the date specified in the request for the grant of the new tenancy.

If agreement is made for the grant of a new tenancy in accordance with the requirements of section 28, this will end the continuation tenancy on the date when the new tenancy is to commence (see Chapter 13).

It will be appreciated from the above outline that although a section 25 notice will specify a date on which the current tenancy is to terminate (see 6.4) and a section 26 request will specify a date on which the new tenancy is to start, having the effect that the termination date for the current tenancy will be the day before that start date (see 7.3), the current tenancy will in fact continue beyond the relevant termination date if the tenant remains in occupation and either an application to the court is made by the termination date or the landlord agrees to extend the time limit for the court application (as described in Chapters 4 to 10). In the absence of those protective steps, the tenant's continued right to occupy the premises will end on the relevant termination date.

The notice procedures under sections 25, 26 and 27 of the Act are discussed separately in Chapters 6 to 10. Agreement as to a new tenancy is discussed in Chapter 13.

If the tenant does not have the protection of the Act because he is not in occupation of any part of the premises, his tenancy will end on the expiry of the term of his lease or, if applicable, on the expiry of a valid contractual notice to quit given by the landlord (see 4.5).

If the tenant is in occupation of the premises for the purposes of his business but nevertheless does not have the protection of the Act for some reason, such as that his tenancy validly contains an agreement excluding sections 24 to 28 (see Chapter 3), his tenancy will end on the expiry of its term or, if applicable, on the expiry of a valid break notice or contractual notice to quit given by either the landlord or the tenant under the terms of the tenancy. If the tenant in that situation nevertheless remains in occupation, there may be a number of legal consequences. If he remains in occupation while he negotiates with the landlord for a new lease which they have agreed is also to exclude protection, he may be treated as a tenant-at-will and remain unprotected by the Act. Alternatively, the landlord might expressly agree with the tenant that the tenant may remain in occupation temporarily as a tenant-at-will; if that is a bona fide arrangement, it may be effective (and, unless anything to the contrary is agreed, the tenant may be liable to pay a full market rent during that period and not just the same rent he was paying previously: *Dean & Chapter of Canterbury* v *Whitbread* [1995] 1 EGLR 82, ChD). However, if there are no ongoing negotiations for the tenant to be given a new tenancy excluded from the protection of the Act, and the landlord nevertheless accepts rent from the tenant for periods after his excluded tenancy has ended, the situation may be treated as the creation of a new periodic tenancy within the protection of the Act. These possibilities were further discussed at 2.2 and 2.3.

4.2 The continuation tenancy

Where the tenant who has the protection of the Act remains in occupation after the expiry of the term of his lease (often referred to as his "contractual tenancy"), it will be assumed that he does so in exercise of the continuance of that tenancy under section 24 of the Act (as outlined at 4.1), unless there is some clear indication that he remains there on some other basis (*Winter* v *Mobil Oil Co Ltd* (1975) 119 Sol J 398). The period of continuance, under section 24, of a contractual tenancy is commonly called the "continuation tenancy".

Since section 24 operates by continuing the existing tenancy, all the terms of the tenancy continue to apply, except those relating to the expiry of the tenancy (such as the yielding-up clause) which then take effect at the end of the continuation period and may be affected by the grant of a new tenancy.

The contractual level of rent will continue to be payable during the continuation period, subject to the provisions for interim rent (see Chapter 11). Ordinarily a rent review clause in the lease will not be interpreted to require rent reviews to take place during the continuation period, unless the wording of the lease clearly indicates such an intention, for example by defining the term to include any statutory continuation and by providing for reviews by reference to a periodic basis throughout the term, rather than specifying actual dates (*Willison* v *Cheverell Estates Ltd* [1996] 1 EGLR 113). On the other hand, a turnover-based rent arrangement applying to the contractual tenancy might more readily be treated as continuing to apply during a continuation tenancy (*Berthon Boat Co Ltd* v *Hood Sailmakers Ltd* [2000] 1 EGLR 39).

If the continuation tenancy is running, there is no need for the landlord to stop collecting rent and he should not do so. If he applies for an interim rent (see Chapter 11), he should demand the old rent on an "on account" basis until the interim rent has been determined.

The continuation tenancy under section 24 is in respect of all the property included in the existing tenancy, and is not limited to the holding occupied by the tenant. Thus where a tenant occupies part of the property and has sublet other parts, during the continuation period the subtenants remain the tenants of the tenant, who himself remains the tenant of the whole property.

Since it is the "tenancy" (the relationship between the present landlord and the present tenant) which continues under section 24, rather than the contractual term of the lease, the effect is that primarily only the present tenant is liable to pay the rent and perform the tenant's covenants under the lease during the continuation period (*City of London Corporation* v *Fell* [1993] 2 EGLR 131, CA). Even if the tenant is an assignee and his lease had been granted before 1996 and was therefore not a "new tenancy" for the purpose of the Landlord and Tenant (Covenants) Act 1995), the continuing liability of previous tenants under the lease by virtue of privity of contract will not automatically make them liable in respect of this continuation period. However, the wording of their covenants with the landlord may create that ongoing liability, for example where they covenanted in respect of the "term" and that expression was defined as including any continuation period (*Herbert Duncan Ltd* v *Cluttons* [1993] 1 EGLR 93, CA).

Similarly, since it is the relationship between just the landlord and the tenant which is continued by section 24, if the landlord has the benefit of a guarantee from a guarantor (or surety) for the tenant's liabilities under the contractual tenancy, that guarantee will not apply to the tenant's liabilities under the continuation tenancy unless the continuation tenancy is encompassed by the guarantor's covenant on a proper interpretation of the covenant in the context of any related provisions of the contractual tenancy documents (*Junction Estates Ltd* v *Cope* (1974) 232 EG 355).

4.3 Termination by surrender

An actual surrender by the tenant of his tenancy is effective without having to comply with any particular provisions of the Act, provided that the surrender takes immediate effect (*Tarjomani* v *Panther Securities Ltd* (1983) 46 P&CR 32). On the other hand, an agreement that the tenancy will be surrendered at a later time (an agreement to surrender) will be void where the tenancy is protected by the Act unless the agreement is made following compliance with the relevant validation procedures under the Act (*Joseph* v *Joseph* [1966] 3 WLR 631).

This equally applies to an agreement to surrender which arises when a landlord accepts an offer of surrender which a tenant makes when complying with a surrender-back clause in a lease (*Alnatt London Properties Ltd* v *Newton* [1983] 1 EGLR 1, CA; and see 3.8).

Before the 2004 amendments, there was a rule that a surrender could not be made by a written instrument which had been executed before, or was executed pursuant to an agreement which had been made before, the tenant had been in occupation for one month (section 24(2)(b)). This old provision was designed to avoid landlords getting tenants to execute deeds of surrender before letting them into occupation, but it has not been relevant since 1969, when section 38 was amended to allow contracting out, and it was eventually abolished by the RRO.

The procedure for ensuring the effectiveness of an agreement to be entered into by the tenant with his landlord that he will, at a later time, surrender his tenancy, is set out at 3.7.

When a surrender is completed, it ends the contractual term of the tenant's lease and the tenant has no rights of continuation or renewal under the Act. However, if the tenant has sublet all or any part of the premises, the contractual term of the subtenancy will continue unaffected by the surrender of the lease out of which it was granted and the subtenant will become the direct tenant of the superior landlord who accepted the surrender (see *Pennell* v *Payne* [1995] 1 EGLR 6, CA; *Barratt* v *Morgan* [2000] 1 EGLR 8, HL; *PW & Co Ltd* v *Milton Gate Investments Ltd* [2004] 3 EGLR 103, ChD). For the different effect on a subtenant where a superior lease ends by the exercise of a break clause, see 4.5 below.

4.4 Termination by forfeiture

The Act does not prevent the tenant's lease from being forfeited by the landlord (section 24(2)). The landlord may do so in the event of non-payment of rent or breach of covenant by the tenant. There is also an ancient common law rule that a sublease will be ended by forfeiture of a superior lease and this is also allowed by the Act.

However, while the tenant is validly seeking relief from forfeiture, the tenancy remains capable of being continued by section 24 (*Meadows* v *Clerical Medical & General Life Assurance Society* [1980] 2 EGLR 63).

There is a large body of law on the subject of landlord's rights of forfeiture and the rights of tenants and subtenants (and their mortgagees) to seek relief from the court against forfeiture. More detail on this subject is outside the scope of this book.

4.5 Termination by notices to quit and break notices

A periodic tenancy, by its nature, can be ended by a notice to quit given by either party. The period of notice will depend on the periodic nature of the tenancy, on which there are common law rules which are outside the scope of this book.

Sometimes a fixed term lease will contain provisions for the term to be ended, before its expiry, by the service of a break notice which may be given by a particular party or, occasionally, by either party. The break may be stated either to take effect only on one or more specified fixed dates or at any time after a specified date. Sometimes the provisions require specified conditions to be fulfilled for a break notice to be served, such as the landlord intending to redevelop the property. In other cases the provisions of the lease allow a break notice to be given without conditions.

Service of a proper notice to quit will end the contractual period of a periodic tenancy. In the case of a break clause in a lease which otherwise grants a fixed contractual term, unless the clause requires the satisfaction of a condition which has not in fact been fulfilled, the service of the break notice in accordance with the terms of the lease will, when it takes effect, operate to end the contractual term of the lease. Where the tenant has sublet all or any part of the premises for a term extending past the break date, the contractual term of the subtenancy will end automatically when the superior lease ends as a result of a break notice, whether or not the sublease itself contains a break provision This applies whether the break notice under the superior lease is given in pursuance of a landlord's break clause or a tenant's break clause (*Pennell* v *Payne* [1995] 1 EGLR 6, CA; *Barratt* v *Morgan* [2000] 1 EGLR 8, HL; *PW & Co Ltd* v *Milton Gate Investments Ltd* [2004] 3 EGLR 103, ChD). The service of a break notice by a tenant is not the same as him surrendering his lease; a surrender takes place under a consensual arrangement made between the landlord and the tenant after the grant of the lease and has a different effect on subtenants (see 4.3 above).

If the break notice is served by the tenant, it will end not only the contractual tenancy but also, so far as that tenant is concerned, the protection of the Act. This is because section 24(2) provides for the tenancy to end by a "notice to quit given by the tenant". "Notice to quit" is defined in section 69(1) as a notice to terminate a tenancy "whether a periodic tenancy or a tenancy for a term of years certain" and therefore encompasses the exercise of a break clause in a lease as well as the termination of a periodic tenancy. Accordingly, where the tenant gives such a notice, it will end both his contractual term and his tenancy, so there will be no continuation under section 24 after the notice takes effect. Further, in that situation the tenant cannot give the landlord a section 26 request claiming a new tenancy (*Garston* v *Scottish Widows Fund* [1998] 2 EGLR 73, CA); if the tenant could do so, he could use a break clause effectively to bring about a downward rent review.

If the tenant serves a break notice to end his lease where he has sublet all or any part of the premises for a term extending past the break date, the contractual term of the subtenancy will (as mentioned above) end when the contractual term of the superior lease is ended by the notice, but the subtenant will have, as against the superior landlord, rights of continuation and renewal under the Act for the premises sublet to him, unless his subtenancy has been excluded from protection (*PW & Co Ltd* v *Milton Gate Investments Ltd* [2004] 3 EGLR 103, ChD). In that situation the landlord can serve a section 25 notice on the subtenant or the subtenant can serve a section 26 request for a new tenancy on the landlord. Alternatively the subtenant could use the opportunity to vacate on or before the break date (see 4.6). If the landlord and the subtenant do nothing and the subtenant remains in possession, there will be a continuation of his subtenancy.

If the break clause in a lease is drafted erroneously and contemplates that subleases will not end when a break notice ends the contractual term of the lease, that cannot have the effect in law of avoiding the termination of the contractual terms of subleases, since one cannot contract out of the rule in *Pennell* v *Payne* mentioned above. However, in exceptional circumstances it may be that, by an equitable doctrine such as estoppel or convention, the landlord and the subtenant are prevented from denying the continued existence of the contractual term of a sublease until it ends for some different reason (*PW & Co Ltd* v *Milton Gate Investments Ltd* [2004] 3 EGLR 103, ChD).

The position of tenants under the Act is different in cases where a break notice is served by the landlord, rather than by the tenant. If, at the time when the notice is served, the tenant or a subtenant has the protection of the Act, the landlord would have to serve a section 25 notice (as well as, ideally, a separate break notice — see 4.5 and 6.6) on the tenant and/or subtenant in order to end both the contractual term held by that person and his tenancy under the Act (*Morrison Holdings Ltd* v *Manders Property (Wolverhampton) Ltd* [1976] 1 EGLR 70).

If in this situation the landlord serves only a contractual break notice, it will end the contractual term but will not end not the tenancy (*Weinbergs Weatherproofs Ltd* v *Radcliffe Paper Mills Co Ltd* (1957) 170 EG 739). In such cases the tenancy will continue by virtue of the section 24 of the Act and the tenant and/or subtenant may remain in occupation (*Scholl Manufacturing Co Ltd* v *Clifton (Slim-Line) Ltd* [1967] Ch 41; *Willison* v *Cheverell Estates Ltd* [1996] 1 EGLR 113, CA). In such cases the tenant or subtenant could, if he wished, serve a section 26 request claiming the grant of a new lease. Alternatively the landlord could serve a section 25 notice once he realises that serving the break notice will only end the contractual term and not the tenancy; however, because of the six month minimum period for section 25 notices, the termination date for the tenancy may then have to be later than the contractual break date.

Where a break notice is served by a landlord, it is open to the tenant or subtenant to vacate the premises on or before the break date, whether or not the landlord has served a section 25 notice, if the tenant prefers to do that rather than claim the protection of the Act (see 4.6).

If the landlord does not have a ground available to him under section 30 on which he can oppose the grant of a new tenancy (see Chapter 8), he can still serve a section 25 notice in relation to a break date but the notice would have to state that he was not opposed to the grant of a new tenancy and he might have to grant the tenant a new lease to replace the lease that has been terminated. Nevertheless that might be advantageous to the landlord where the break date occurs substantially before the next rent review date (or term expiry date) and the rent that would be fixed under section 34 (see 12.6) under a new tenancy will be higher than the passing rent, or where the tenant is not in occupation of the whole premises and the landlord wants to take back the part which the tenant does not occupy.

If the tenant or subtenant is not protected by the Act at the time when the landlord serves the break notice, for example because he is then not in occupation of the premises, he cannot obtain the protection of the Act afterwards even if he takes up occupation before the notice expires (section 24(3)(b)).

4.6 Tenant not wanting a continuation

If a tenant protected by the Act does not want his contractual term to be extended by section 24 or by any of the other procedures of the Act (see Chapters 4 and 5), he has a choice of two actions. He can vacate the premises by the contractual term date without having to notify the landlord (section 27(1A)). Alternatively, if the tenant holds a fixed-term tenancy which has at least three months unexpired, he can give the landlord a notice under section 27(1) stating that the tenancy will not continue beyond the term expiry date. The effect of vacating or giving the notice will be to exclude section 24 from operating to continue the tenancy beyond the expiry of the contractual term. The form of section 27(1) notice is not prescribed under the Act.

A notice under section 27(1) can be given, or the tenant can simply vacate, even where the tenant has been given a section 25 notice or has given a section 26 request which creates a termination date falling after the end of the contractual term, but to take advantage of this rule the tenant must vacate by the end of the contractual term and not the later date set by the section 25 notice or section 26 request.

The tenant's right to vacate by the term expiry date without giving notice can put the landlord into a difficult position, especially since the deadline for the tenant to apply to the court for a new tenancy similarly does not arise until the expiry date of the section 25 notice or section 26 request. During the intervening period after the service of the initial notice or request, the landlord may not know whether the tenant intends to renew or vacate. Ideally there should be dialogue between the parties, but if the tenant is not forthcoming the only method by which the landlord can force the issue appears to be for the landlord to make the court application (discussed in Chapter 10) and see how the tenant responds.

If the current lease allows him to do so, the landlord might erect a letting board outside the premises, either with the genuine wish to attract new tenants or simply to provoke the current tenant into declaring his intentions.

Even if there has already been an application made to the court under section 24(1) or section 29(2), termination under the provisions of section 27(1A) will operate and will seemingly not depend on the cessation of the court proceedings. This is because the provisions of section 64, which create a further continuance where proceedings are on foot (see Chapter 10) are expressed to apply to Part II of the Act only where, apart from section 64, the tenancy will end by the operation of a termination or section 26 request. Where the tenancy ends due to some other event (in this case, the tenant vacating before the term date in reliance on section 27(1A)), the termination process set by the section 25 notice or section 26 request will have ceased to operate and therefore section 64 will not be brought into play (see *Single Horse Properties Ltd* v *Surrey County Council* [2002] 2 EGLR 43, CA).

Where the property comprised in the tenancy is a house (see Chapter 2), the tenant cannot validly serve a section 27(1) notice during the currency of a claim he has made under the Leasehold Reform Act 1967 to acquire the freehold or an extended lease of the house (1967 Act, schedule 3 para 1(2), (3)). Presumably this rule does not prevent him from ending the application of the 1954 Act by simply vacating the house.

4.7 Tenant ending a continuation tenancy

Once a continuation of the contractual tenancy has started, it will not end merely by the tenant vacating the premises (section 27(2)). In order to end the continuation, the tenant must give his immediate landlord a notice under section 27(2) to bring the tenancy to an end. The termination date to be specified in this notice can be any day falling at least three months after the day when the notice is given (prior to the 2004 amendments, this notice had to expire on a quarter day) and the effect of the notice will be to end the tenancy on that date. The form of section 27(2) notice is not prescribed under the Act.

A notice under section 27(2) can be given even where the tenant has been given a section 25 notice or has given a section 26 request with several months still to run, and the section 27(2) notice will operate to end the tenancy on the termination date specified in it even if that is earlier than the termination date created by a section 25 notice or section 26 request.

If there has been an application made to the court under section 24(1) or section 29(2), it appears that a notice given under section 27(2) will apply but termination will depend on the cessation of the court proceedings. This is because the provisions of section 64, which create a further continuance where proceedings are on foot (see Chapter 10) are expressed to apply where, apart from section 64, the tenancy will end by the operation of a notice which might be a landlord's section 25 notice or a tenant's section 27 notice — or a section 26 request. In such cases, the effect of section 64 is that the termination date arising under the notice or request will be deferred until the expiration of three months from the date on which the court application is "finally disposed of". The final disposal of court applications is considered in detail at 10.23.

In cases where the tenant had paid rent in advance and serves a section 27(2) notice, he can recover from the landlord an apportioned part of the rent referable to the period after the section 27(2) notice expires.

4.8 Tenant losing protection during a continuation tenancy

If the tenant loses the protection of the Act during a continuation tenancy by ceasing to continue in occupation of at least some part of the premises for the purpose of his business, the continuation tenancy will not in fact end immediately. If it is a continuation of a tenancy which was created by a fixed term lease, the continuation tenancy can be terminated by the landlord giving not less than three nor more than six months' written notice to the tenant, or in accordance with any different termination provisions in the lease (section 24(3)(a)). It must be remembered that if the tenant has vacated due to matters beyond his control and wishes to return to full occupation, he may not in fact have lost the protection of the Act and in such cases the landlord cannot serve notice under section 24(3)(a) of the Act (see the cases on fire damage at 2.6).

Alternatively, the tenant can end any continuation tenancy by serving on the landlord a three months notice under section 27(2), mentioned above. If he does, and if he has paid rent in advance, he can recover from the landlord a apportioned part of the rent referable to the period after the section 27(2) notice expires.

As mentioned at 4.1, the tenant may also lose protection where a section 25 notice or section 26 request is served and, by the relevant termination date, he fails to reach agreement with the landlord for a new tenancy and neglects to ensure that an application is made to the court. Where an application to the court is made, special rules under section 64 apply to ascertain the effective termination date (see Chapter 10), except as mentioned at 4.7.

4.9 Premises with telecommunication operator's apparatus

The Telecommunications Code includes provisions under which telecoms operators can acquire, by agreement with the person in possession of land, the right to install, keep, maintain, adjust repair and alter "electronic communications apparatus" on the land; compulsory powers for telecoms operators to acquire such rights by court order if they cannot obtain them by negotiation; and, once such rights exist, protection from having to remove the apparatus without a court order given under rules set out in the code.

The code is set out in schedule 2 of the Telecommunications Act 1984, as amended by the Communications Act 2003 (the 2003 Act). The following is only a brief outline of a complex set of provisions. The code is based on a statutory principle that no person should unreasonably be denied access to a telecoms system. If a lease to a telecoms operator is contracted out of sections 24 to 28 of the 1954 Act (see 2.16 and Chapter 3), this does not exclude the right of a telecoms operator to claim protection under the code where the landlord seeks possession from him and requires removal of the apparatus from the premises.

"Electronic communications apparatus" is widely defined and includes any equipment, machinery or device and any wire or cable (and the casing or coating for any wire or cable) which is designed or adapted for use in connection with the provision of an electronic communications network or for sending or receiving communications or other signals transmitted by such a network. It also includes wires, cables, tubes, pipes and similar things designed or adapted for use in connection with the provision of an electronic communications network or service, and any conduit carrying any such apparatus, any structure on which it is installed, and any pole or other thing from which it is suspended or which supports it (code para 1(1); 2003 Act section 405).

An agreement in writing granting rights in respect of telecoms equipment may be made under para

2 of the code. The rights may simply be to execute works on the property, or to keep or install the apparatus on the property, or to enter the land to inspect apparatus located on that or other property.

Generally, an owner or lessee will not be bound by such an agreement unless he granted it himself as the occupier or, if the land is vacant, as being the person entitled to possession; or he agreed in writing to be bound by it; or the agreement relates to the provision of a telecoms service to the occupier; or he is the successor to a freeholder or leaseholder who entered into the agreement and the agreement did not contain a provision that it will not bind that person's successors in title. So, for example, the landlord of premises will not be bound by an agreement granted by his tenant unless he agreed in writing to be bound by it.

An owner or lessee who is not at a point in time bound by such an agreement, either because he was never bound by it or because the agreement has ended, will have the right to require the operator to remove the equipment and restore the land (code para 4(2)). In the case where the operator has a lease of the premises on which he has installed the apparatus, this might arise when his tenancy is properly ended. However, even if the tenancy is properly ended — and even if this is at the end of a contracted-out lease (see Chapter 3) — the owner's right to have the operator cease use of the apparatus and remove it is subject to restrictions set out in para 21 of the code, summarised below. For this reason, para 4(4) of the code provides that when there is a grant of rights under para 2 of the code, "compensation" is payable by the operator to such a person for any depreciation in the value of his interest in the land arising by virtue of the fact that he will be bound by the operator's rights under para 21 of the code at the time when he would otherwise become entitled to take occupation. The amount of compensation, if not agreed, is to be determined by the Lands Tribunal under the Land Compensation Act procedure.

If an agreement under para 2 of the code cannot be negotiated, an operator can apply to the court for an order allowing him to install and retain equipment in a particular place (code, para 5). Such an order will only be made if the loss to the landowner can be adequately compensated for by money or is outweighed by the benefit accruing to the persons who would obtain access to the network or system if the order were made. The court must have regard to all the circumstances and to the statutory principle that no person should unreasonably be denied access to a telecoms network or system. The order is to be made on terms and conditions so that the least loss and damage is caused in the exercise of the rights. The amount of compensation is to be determined by the court under para 7 of the code and will presumably depend on the value of the land (or part of the building) and the other uses to which it might otherwise be put and how they are prejudiced. Similar provisions apply under para 6 of the code in certain cases where the equipment has already been installed on the land. It is believed to be quite rare for orders to be made under para 5 of the code, since telecoms operators prefer to deal with these matters by negotiation.

Para 21 of the code applies where the owner of property has a legal right to require a telecoms operator's equipment to be removed from it. This could apply on the termination of an agreement made under para 2 of the code or where the owner or lessee has this right because he is not bound by such an agreement. No such right to secure removal can be enforced without first serving notice on the operator, who may within 28 days serve a counter-notice which has the effect that the right to secure removal cannot be exercised without a court order. The operator's counter-notice must state if the operator contends that there is in fact no right to require the removal, and/or specify the steps which the operator proposes to take to secure the right to keep the equipment in place. If there is no basis on which the operator can successfully challenge the legal right to require removal of the equipment, the operator can nevertheless seek to demonstrate that he would be able to secure compulsorily the right to retain the equipment in its present place under the provisions of paras 5 and 6, mentioned above. The burden of proof imposed in this way on the operator is quite onerous. If he fails, the court order is granted to the land owner and the removal of the equipment will be at the operator's cost.

Para 20 of the code applies whenever the owner of the property does not have a legal right to require the removal of telecoms apparatus or have it altered, but wants to make an improvement to his property and this would be impeded by the apparatus if it were not altered. For this purpose, improvement includes a change of use as well as work of development. This situation might arise while an agreement made under para 2 of the code exists which is binding on him, or where the rights were ordered by the court under para 3 of the code. In this situation, the owner wishing to carry out an improvement of his land may apply for the apparatus to be altered. He must serve a notice on the operator, who within 28 days may serve a counter-notice which has the effect that the alteration need not be carried out by the operator unless the landowner obtains a court order requiring that to happen. The criteria for the landowner obtaining the court order are quite stringent; he must satisfy the court that the alteration to the apparatus is necessary for the purpose of carrying out the improvement and that the alteration will not substantially interfere with any service operated over the network using the apparatus. The court shall not make the order unless the operator would have all appropriate rights for this purpose including rights which the court would grant him under para 5. If the court order is granted, the alteration of the equipment is to be at the landowner's cost.

It is not certain whether the courts will enforce an indemnity given by the operator to a land owner in a para 2 agreement, undertaking to compensate the owner for loss suffered as a result of non-removal of apparatus under the provisions of para 21, if it purports to go beyond the compensation arising under para 2(4) of the code.

4.10 Stamp duty land tax on continuation tenancies

A special rule, set out in schedule 17A of the Finance Act 2003, applies where the lease which is being continued is one which fell within the regime of stamp duty land tax (SDLT). It may result in SDLT becoming payable in respect of the continuation tenancy. The rule does not apply where the tenancy which is being continued had been granted at a time when the old stamp duty regime applied, before 1 December 2003. It applies to the continuation of any lease which was granted on or after 1 December 2003, whether or not SDLT shall actually have been payable on that lease — it applies even if the term and rent of the lease had resulted in the lease being below the threshold for SDLT to be payable. As soon as the tenant starts holding over after the end of such a lease, that lease is immediately treated for SDLT as if it had been granted for a term one year longer than its actual term. This is sometimes colloquially called the "growing" lease.

SDLT is assessed on the "net present value" (NPV) of the rent payable throughout a lease, under a complicated formula set out in the legislation which takes account of the length of the lease and the amount of rent. In the case of non-residential premises, at the date of this book the threshold below which no SDLT is payable is where the NPV of the lease does not exceed £150,000.

As soon as the continuation tenancy begins, the tenant must recalculate the NPV of the lease on the basis of the whole of the notionally lengthened term, assess SDLT on that basis, and pay the amount by which that SDLT exceeds the amount of SDLT (if any) that had been payable when the lease was originally granted. The rates of SDLT are those applicable to the original lease.

For example, if a lease was granted on 1 January 2004 for a term of five years at a rent of £150,000 per year, its NPV would have been calculated as £677,257 giving rise to SDLT of £5,272. If the tenant begins holding over on 1 January 2009, the NPV is then to be recalculated as if the lease had been granted for a six year term, which gives an NPV of £799,282 on which SDLT would be £6,492, and so an additional SDLT payment of £1,220 becomes due.

If the continuation tenancy lasts longer than a year, this has to be repeated immediately after the end of that year, when further SDLT is to be calculated, giving credit for any SDLT previously paid under this rule as well as any paid on the original grant of the lease. This process is repeated every time the continuation tenancy runs on for a further year.

The tenant is under a duty to submit to the Inland Revenue a return, on forms SDLT1 and SDLT4, within 30 days of the start of each relevant year of the continuation tenancy and to pay any SDLT then payable. However, a return does not have to be submitted if no SDLT is payable if, on making the calculation, the NPV remains below the threshold, but only in those cases where the original fixed term of the lease was less than seven years, since that will not be a "notifiable" transaction (Finance Act 2003 section 77(2)(b)(ii), schedule 17A para 3(5)).

If the tenant pays SDLT in respect of a continuation year and he is granted an actual new lease of the premises before the end of that year, HM Revenue and Customs seems to accept that, notwithstanding the rather uncertain application of the precise wording of the statutory provisions, the tenant can obtain relief for the overlap by calculating the SDLT on his new lease as if the rent payable under it in respect of the overlap period was reduced by the amount of rent that would have been payable under the continuation tenancy in respect of that period (Finance Act 2003, schedule 17A para 9).

As for the SDLT liability in respect of interim rent, see 11.6.

4.11 Giving possession on termination

In most cases, where the existing tenancy ends under the above rules (or, following a court application, as mentioned at 10.23) and a new tenancy is not being granted, the tenant must give up possession to the landlord and must comply with any covenants in his lease as to the condition of the property at that time (often contained in a "yielding up" covenant and in the repairing and decorating covenants).

Where part of the property has been sublet by the tenant, the question whether that part must be vacated depends on the circumstances. If part is sublet to a tenant occupying for business purposes and the subtenant is protected by the Act, the landlord should have served a section 25 notice on the subtenant and dealt with the subtenant's rights under the Act in tandem with the tenant's (see Chapter 9).

Where a tenant having a business lease of premises which include a residential unit has sublet the residential unit on a tenancy protected by the Rent Act 1977, the subtenancy will be binding on the superior landlord at the end of the business lease (*Wellcome Trust* v *Hammad; Ebied* v *Hopkins; Church Commissioners for England* v *Baines* [1998] 1 EGLR 73, CA).

4.12 Dilapidations

Where no new lease is granted and the landlord retakes possession of the premises, he will ordinarily be able to claim damages from the outgoing tenant if the premises are handed over in a poorer state of repair than that required of the tenant under the terms of the lease which has terminated. The assessment of such damages is a complex question and is outside the scope of this book, but generally the landlord cannot recover in damages more than the amount by which the value of his reversion is diminished by the disrepair (Landlord and Tenant Act 1927, section 18).

As for the case where a subtenant remains in possession, see 9.3.

Preliminary Matters

5

5.1 Identifying the "landlord"

The person who has the functions and powers of the "landlord" under the Act in relation to a particular occupying tenant is not necessarily the tenant's immediate landlord, if that person is not the freeholder but only has a leasehold interest in the premises. Because the essence of the Act is to give the tenant a right to be granted a new tenancy, the landlord has to be someone (called the "competent landlord" in schedule 6 of the Act) who has more than just a very short or nominal reversion to the tenant's current tenancy. Since a landlord's section 25 notice (see Chapter 6) can normally specify a termination date up to 12 months ahead, the Act provides that a leaseholder who is a landlord of an occupying tenant must himself have a tenancy which will last for at least 14 months in order to be the competent landlord of the occupying tenant for the purposes of the Act (section 44(1)). If he does not, a superior landlord will be the competent landlord of the occupying tenant.

Since a tenant who has sublet part of his premises but remains in occupation of other parts of the premises for the purpose of his business will himself be protected by the Act in relation to his entire tenancy (see 4.1), even when the contractual term of his lease has less than 14 months unexpired, he will not cease to be the competent landlord of his subtenant under section 44 until he has been given a section 25 notice by his own landlord or has given his own landlord a section 26 request (*Bowes-Lyon* v *Green* [1963] AC 420; *Cornish* v *Brook Green Laundry Ltd* (1959) 173 EG 307, CA). The potential for his tenancy to extend beyond the termination date of the section 25 notice or section 26 request, by virtue of section 36(2) or section 64 of the Act (see 10.23), does not result in the tenant remaining a competent landlord (section 44(1)(b)). The desirability that a leaseholder should serve notices on his subtenants before the shortness of his own term loses him the competency to do so is discussed further at 9.2.

The tenant may, depending on the terms of his documentation, be treated as if his tenancy has more than 14 months unexpired if he has a legal right to be granted a new lease of the premises under an agreement for lease or under an option for renewal which he has exercised, even if the new lease has not yet been granted to him (*Shelley* v *United Artists Corporation Ltd* [1990] 1 EGLR 103, CA).

Where there is a hierarchy of leases, each leaseholder up the chain, starting with the occupying tenant's immediate landlord, is tested against the 14 month rule to ascertain if he is the competent landlord of the occupying tenant. If there is no leaseholder having a tenancy which will run for at least 14 months, the freeholder will be the competent landlord of the occupying tenant and in that situation the occupying tenant's immediate landlord will simply have the limited function of the "mesne landlord" under the Act (see Chapter 9).

Where parts of premises in different ownership (freehold or leasehold) are let together under a single tenancy, any steps that the different landlords wish to take under the Act can only be taken collectively and not individually (section 44(1A)).

In this book, as in the Act, references to the "landlord" generally mean the competent landlord.

It is possible that a person who is the competent landlord of the occupying tenant at the time when he gives him a section 25 notice or receives a section 26 request will cease to be his competent landlord at some time after that date. Take the case of a tenant who holds a lease having, say, 10 years unexpired but which is subject to a break clause under which the landlord may serve six months' notice to end the lease at any time. If the tenant sublets the whole property (so that he ceases to have the protection of the Act himself), he will be the competent landlord of his subtenant unless and until he receives a break notice from his own landlord or until his own lease has less than 14 months unexpired. The tenant can give the subtenant a section 25 notice while he is still the subtenant's competent landlord. If he does so, say six or seven months before the expiry date of the sublease, but after doing so he receives a break notice from his own landlord, he will then cease to be the competent landlord of the subtenant.

From the moment that a superior landlord becomes the competent landlord of the subtenant, any dealings under the Act must be between the superior landlord and the subtenant. If the subtenant continues to act as if his immediate landlord is still his "landlord" for the purpose of the Act and fails to join the superior landlord as defendant in his proceedings for a new tenancy (see Chapter 10), he will lose his rights of renewal (*Rene Claro (Haute Coiffure) Ltd* v *Halle Concerts Society* (1969) 20 P&CR 378, CA).

5.2 Preliminary investigations

It is vital for a landlord to be able to find out details of all persons who are occupying the premises as tenants (including subtenants) inside the Act, since he needs to serve section 25 notices on all of them if he is to be certain of either getting the full benefit of lease renewals or obtaining possession if he is opposing renewal. If he wants to oppose lease renewals on the ground of, say, redevelopment (see Chapter 8) but fails to serve a section 25 notice on a protected subtenant, that could delay his redevelopment, at huge cost. Inspection alone may not readily reveal the presence of sub-tenants, so accurate information from the tenant is vital.

Equally it is essential that the tenant can identify the competent landlord who has to be named as defendant in any renewal proceedings (see 5.1). If he addresses the proceedings to the wrong person and does not have time to correct it before the deadline runs out, he may lose his right of renewal, again possibly suffering serious loss.

5.3 Investigations by the landlord

Before a section 25 notice is served (see Chapter 5), the landlord should verify the identity, status and correct names and addresses of all tenants and subtenants, as mentioned at 5.2. The landlord must consider which of the occupiers are protected by the Act (see Chapter 2) and may have to be served with a section 25 notice. This should be done in the first instance by inspection, enquiry of occupiers, checking leases and ancillary documents such as licences, making company searches, and serving notices under section 40 of the Act (see 5.5). This can be done up to two years before the expiry dates of the tenants' leases, but should in practice be carried out two or three months before the date on which the landlord intends to serve the section 25 notices and the information should be rechecked, so far as possible, immediately prior to service.

Where the tenant's lease is registered at the Land Registry (see 2.2), the landlord should obtain an up to date copy of the registered title and check the identity of the registered owner of the lease.

If the tenant is a UK company, a search at Companies House would confirm the current name of the company and the address of its registered office.

The landlord should then consider which occupiers are protected by the Act (see Chapter 2) and need to be served with section 25 notices, and he should endeavour to identify the extent of each such tenant's holding (see 16.1).

Inspection of the property should include consideration of whether there is disrepair due to the tenant's default and whether there have been any breaches of other covenants (eg as to alterations, user or alienation). The tenant's record of payment of rent should also be checked for any persistent delays in paying rent. This will enable a decision to be taken whether the landlord should oppose a renewal on ground (a), (b) and (c) of section 30(1), being disrepair, persistent delays in paying rent, or other substantial breaches or mismanagement (see Chapter 8).

A landlord should be cautious about serving a section 25 notice when it is not necessary or desirable. For example, where a tenant or subtenant is currently paying rent equal to or above market level and there is no benefit to the landlord's reversion to ensure that the property is let for any particular term, the landlord might sensibly decide to let the tenant or subtenant hold over under a continuation tenancy for the moment.

A landlord should also be cautious of specifying a ground of opposition if he knows that the ground has no validity at all and that it is in effect being put forward fraudulently. While it had been said that specifying a ground in a notice is merely giving advance warning (in the nature of court pleadings) that an issue is to be argued in court at a later date, Lord Denning in *Betty's Cafes Ltd* v *Phillips Furnishing Stores Ltd* [1959] AC 20 said: "If the notice had been a dishonest notice in which the landlords had fraudulently misrepresented their intention — or, I would add, if there had been a material misrepresentation in it — I should have thought it would be a bad notice", and in a subsequent case relating to an agricultural tenancy it was held by the Court of Appeal that a notice to quit founded on alleged breaches of covenant was invalid and of no effect since the landlord knew that the breaches had not occurred (*Stradbroke (Earl of)* v *Mitchell* [1990] 2 EGLR 5). However, the *Betty's Cafes* case related to a landlord's counter-notice to a section 26 request for a new tenancy, and section 26(4) requires the landlord to state the ground on which he "will oppose" the tenant's application. That case was distinguished in *Sun Life Assurance plc* v *Thales Tracs Ltd* [2001] 2 EGLR 57 in which the Court of Appeal held that a section 26 request for a new tenancy was merely a request, albeit with a proposal for the terms of the tenancy, and was valid whether or not the tenant genuinely intended to take up a new tenancy if one was available. The question whether the validity of a section 25 notice opposing renewal can be challenged if the landlord's professed ground of opposition is not genuine remains to be decided.

5.4 Forfeiture and dilapidation notices and schedules

If the landlord seriously wishes to obtain possession from the tenant on grounds of tenant's breach of covenant, delaying action until the last year of the tenancy is ill advised. Any appropriate schedule of dilapidations and/or notice under section 146 of the Law of Property Act 1925 should be served well before the intended date for service of the section 25 notice. This is because the court will consider the actions of the parties when considering whether to refuse the grant of a new lease under grounds (a), (b) or (c) of section 30 (see 8.2, 8.3 and 8.4). If the landlord has taken no steps to enforce the relevant covenant during the term, the court is likely to believe that the breach is of no importance to the landlord. However, if the landlord has been actively attempting to enforce the covenants against the tenant and the tenant has failed to comply, this will be telling against the tenant under those grounds of opposition.

Forfeiture of the tenant's current tenancy on the grounds of his breach of repairing covenants, and on the grounds of any other breaches of covenant that the tenant may have committed, should be considered, since an effective forfeiture that would end the tenant's tenancy and give him no rights of renewal (see 4.1). If this is appropriate, a claim for forfeiture should be issued in the courts, preferably before the landlord serves any section 25 notice. The reason is that the right to forfeit for a breach of covenant which is in the "once-and-for-all" category (such as the making of unauthorised alterations) can be lost ("waived") by the landlord or his representative carrying out any act which treats the tenancy as continuing, and serving a section 25 notice may be such an act (see *Barglabasi* v *Deedmethod Ltd* [1991] 2 EGLR 71). Waiver of the right to forfeit does not apply to a "continuing" breach, such as disrepair, so if the only ground of forfeiture is disrepair then the timing of the section 25 notice in relation to the forfeiture may not be crucial. If the tenant obtains relief against forfeiture of his present tenancy, the landlord should in appropriate cases ask the court to impose terms which compensate him for delaying the service of the section 25 notice pending the forfeiture procedure (*cf Soteri* v *Psylides* [1991] 1 EGLR 138, CA). The law relating to forfeiture is complex and more discussion on this topic is outside the scope of this book.

5.5 Landlord's section 40 notice to a tenant

Section 40(1) of the Act enables the landlord to serve notice on a tenant requiring the tenant to give him certain information listed in section 40(2). That information is reflected in the questions on the prescribed notice form, which is form 4 under the 2004 Notices Regulations. The tenant on whom the notice is served may be the landlord's immediate tenant or a subtenant. The section 40 notice may be served at any time after the date two years before the expiry date, or landlord's break date, in the tenant's lease.

The first piece of information asked is whether the tenant occupies all or any part of the premises wholly or partly for the purpose of his business. That information will enable the landlord to assess whether the tenant, at that time, has the protection of the Act by meeting the criteria of occupation and business use (see Chapter 2). Curiously, the notes on form 4 do not alert the tenant to the provisions of the Act by which occupation by certain others — such as group companies, controlled companies, trust beneficiaries, etc — will count as occupation by the tenant (see 2.7 to 2.10). Further, the question on the prescribed form — which follows the wording in section 40 — does not ask the tenant to identify exactly which part or parts of the premises he and his employees occupy, so the landlord at this stage may not be certain as to the extent of the "holding".

The second piece of required information is as to the existence of any sub-tenant and, if so, information as to the premises let to him, the term, the rent, the name of the subtenant, whether (to the best of the tenant's knowledge and belief) the subtenant occupies all or any part of the premises wholly or partly for the purpose of his business (and, if not, the sub-tenant's address), whether the subtenancy is contracted out (see Chapter 3) and whether any notice relating to the subtenancy has been given under section 25 or section 26. That information will enable the landlord to assess whether the subtenant has the protection of the Act by meeting the criteria of occupation and business use (see Chapter 2) and will enable him to serve a separate section 40 notice on the subtenant if he has not already done so.

The third piece of information to be given by the tenant is the name and address of all persons known or believed by the tenant to own an interest in reversion to his tenancy. That will assist the landlord to check the identity of, say, an intermediate leaseholder (the "mesne landlord" under the Act — see Chapter 9) or the owner of another part of the premises where the reversion has been divided up (see 6.1).

The landlord's section 40(1) notice requires the tenant to give his answers in writing within one month beginning with the date of service. On the giving of the notice by the landlord, a legal duty is imposed on the tenant to give the required information within that period (section 40(1), (5)(a)).

A further, quite onerous, duty is imposed on the tenant that if, at any time within the six months from the giving of the original notice, he becomes aware that the information he initially gave in response to it was incorrect or no longer remains correct, he must within one month of becoming aware of it give the corrected information in writing to the landlord (section 40(5)(b)). For example, if in response to the section 40 notice the tenant correctly informed the landlord that he was in occupation of the property but he subsequently vacates it within the following six months, he must inform the landlord.

If the tenant transfers his tenancy at a time when he is under a duty created by a section 40 notice (whether it is the duty to give the initial information within the first month or the duty to give corrected information within the six month period), the duty ceases to bind him only when the landlord is given written notice of the transfer of the tenancy with the name and address of the transferee (section 40A(1)). That will give the landlord an opportunity to serve a fresh section 40(1) notice on the new tenant.

If the landlord transfers his interest after giving a section 40 notice, the tenant may address his response, and any notification of corrections, to the landlord who gave him the notice until the tenant receives written notification of the transfer of the reversion and the name and address of the transferee, after which any response or notification from the tenant must be addressed to the new landlord (section 40A(2)(b)).

Section 40B gives the landlord the right to bring civil proceedings against a tenant for breach of statutory duty if a tenant to whom he gave a section 40 notice fails to supply or update the information in response to the notice. Not only can the court order the tenant to comply with his duties under section 40, but if his default results in loss being suffered by the landlord, the landlord can be awarded damages. Such loss might arise, for example, where the landlord is seeking possession of the property for redevelopment and the tenant does not disclose the existence of a sub-tenant whose presence cannot readily by ascertained by the landlord by other means.

5.6 Tenant's investigations and section 40 notices

For the reasons explained at 5.1 and 5.2, it is important for the tenant to identify his "competent landlord". Section 40(3) of the Act enables a tenant whose current tenancy was granted for a fixed term (whether or not continuing under section 24 — see Chapter 4), or for a fixed term and then from year to year, to serve notice on a reversioner requiring him to give him certain information listed in section 40(4). That information is reflected in the questions on the prescribed notice form, which is form 5 under the 2004 Notices Regulations. The reversioner referred to may be the tenant's immediate landlord or a superior landlord.

The first piece of information is whether the addressee owns the freehold of all or part of the tenant's premises. If he does not, he must also give information, to the best of his knowledge and belief, as to the name and address of his immediate landlord, as to the term of his own tenancy and the earliest date when it could be ended by a landlord's notice to quit, and as to whether any notice has been given in relation to the tenancy under section 25 or section 26. That information will enable the tenant to assess whether the addressee of his notice is the competent landlord of the tenant for the purposes of the Act (see 5.1) and, if the addressee is not, will enable him in appropriate cases to serve a separate section 40 notice on a superior landlord.

The second piece of information is as to the existence of any other reversioner owning all or any part of the premises. That will assist the tenant to check the identity of another owner where the addressee of the notice has transferred all of part of the property to someone else.

The tenant's section 40(3) notice requires the reversioner to respond in writing within one month beginning with the date of service. On the giving of the notice by the tenant, a legal duty is imposed on the reversioner to give the required information within that period (section 40(3), (5)(a)).

A further, quite onerous, duty is imposed on the landlord that if, at any time within the six months from the giving of the original notice, he becomes aware that the information he initially gave in response was incorrect or no longer remains correct, he must within one month of becoming aware of it give the corrected information in writing to the tenant (section 40(5)(b)). For example, if in response to the section 40 notice the landlord correctly informed the tenant that there was no other reversioner, but he subsequently grants an intermediate (overriding) lease of the premises to some other party within the following six months, he must inform the tenant.

If the landlord transfers his reversion at a time when he is under a duty created by a section 40 notice (whether it is the duty to give the initial information within the first month or the duty to give corrected information within the six month period), the duty ceases to bind him only when the tenant is given written notice of the transfer of the reversion with the name and address of the transferee (section 40A(1)). That will give the tenant an opportunity to serve a fresh section 40(3) notice on the new reversioner.

If the tenant transfers his tenancy after giving a section 40 notice, the reversioner may address his response, and any notification of corrections, to the tenant who gave him the notice until the landlord receives written notification of the transfer of the tenancy and the name and address of the transferee, after which any response or notification from the landlord must be addressed to the new tenant (section 40A(2)(b)).

Section 40B gives the tenant the right to bring civil proceedings against the reversioner to whom he gave a section 40 notice for breach of statutory duty if the reversioner fails to supply or update the information in response to the notice. Not only can the court order the landlord to comply with his duties under section 40, but if his default results in loss being suffered by the tenant, the tenant can be awarded damages. Such loss might arise, for example, where the tenant is seeking the grant of a new tenancy but serves his renewal notices or court proceedings on someone who is not the competent landlord due to the tenant having been misled by incorrect information given by the reversioner.

Where a reversioner's interest is mortgaged and the mortgagee is "in possession", which normally means that he is receiving the rent for the premises, the tenant's section 40(3) notice may be addressed to the mortgagee and the duties created by section 40(4) are imposed on the mortgagee.

The tenant may supplement his enquiries under section 40 by checking the landlord's title to the property if it is registered at the Land Registry. This will show the nature of his tenure and the name of the registered owner. If the landlord is a UK company, a search at Companies House would confirm the current name of the company and the address of its registered office.

5.7 Measurement of time-limits

Several provisions of the Act specify critical dates by reference to a number of months either counted from a particular date or before a particular date. An example of the former is section 27(2) which provides for not less than three months' notice to be given under that provision. An example of the latter is section 25(2), which provides that a section 25 notice must not be given less than six months before the termination date specified in it.

When determining these dates, the "corresponding date rule" applies (*Dodds* v *Walker* [1981] 1 WLR 1027, HL). This means that, counting forwards, so long as there are sufficient days in the end month, the number of the day in the starting month will be applied in the end month, so six months from 30 June will be 30 December. Where the end month is shorter than the starting month and has no correspondingly numbered day, the last day of the end month is used, so six months from 31 August will be 28 February in ordinary years and 29 February in leap years.

Where a time-limit is "not less than", as in the case of the minimum of six months in a section 25 notice, specifying a date which is exactly the minimum period ahead of the date of service is valid (*EJ Riley Investments Ltd* v *Eurostile Holdings Ltd* [1985] 2 EGLR 124 CA).

5.8 Methods of serving notices

Section 66(4) of the 1954 Act states that section 23 of the Landlord and Tenant Act 1927 shall apply to the service of notices under the 1954 Act. Section 23 of the 1927 Act provides that notices

> shall be in writing and may be served on the person on whom it is to be served either personally, or by leaving it for him at his last known place of abode in England or Wales, or by sending it through the post in a registered letter addressed to him there.

In these provisions, "last known place of abode" includes the addressee's last known place of business (*Price* v *West London Investment Building Society* [1964] 2 All ER 318) and "registered letter" has been replaced by the recorded delivery service (Recorded Delivery Service Act 1962).

If the addressee is resident abroad but has business interests in England or Wales, service at the business address in England and Wales may suffice (*Italica Holdings Ltd* v *Bayadea* [1985] 1 EGLR 70).

If the notice is to be given to a limited company, service can be in accordance with Companies Act 1985 section 725, by sending it to or leaving it at the registered office of the company (*Robert Baxendale* v *Davstone (Holdings) Ltd; Carobene* v *John Collier Menswear* [1982] 2 EGLR 65). As a precaution, the notice should be served (preferably on the same day) both at the company's business address (in the case of the tenant, usually the premises comprised in the tenancy) and at its registered office.

Section 23 of the 1927 Act also provides that notices to a local or public authority or a statutory or public utility company may be sent addressed to the secretary or other proper officer at the principal office of the authority or company.

The methods of service under section 23 of the 1927 Act mentioned above are not compulsory; any method of actual delivery will suffice if the addressee actually receives the notice. However, using a method of service prescribed by 1927 Act will create an assumption of delivery which cannot be rebutted by evidence that the notice was never actually delivered (*Blunden* v *Frogmore Developments Ltd* [2002] 2 EGLR 29, CA following *Galinski* v *McHugh* [1989] 1 EGLR 109, CA and *Railtrack plc* v *Gojra* [1998] 1 EGLR 59, CA and not following *Lex Service plc* v *Johns* [1990] 1 EGLR 92, CA). Accordingly, a method of service specified in section 23 of the 1927 Act should be utilised whenever possible. Furthermore, the assumed date of service in such cases is the same day as the sender takes the action permitted by section 23. While this is understandable in the case of personal delivery of the notice or leaving it at the premises — in either case there is a chance that the addressee will see it that day — it is anomalous that the rule also has the effect that handing the notice over to the Post Office for recorded delivery creates an assumption that the notice is served on the addressee on the same day. Nevertheless that has been held to be the irrebuttable presumption (*CAWebber (Transport) Ltd* v *Railtrack plc* [2004] 1 EGLR 49, CA; *Beanby Estates* v *Egg Stores (Stamford Hill) Ltd* [2003] 3 EGLR 85, ChD).

Unlike section 196 of the Law of Property Act 1925, the provisions of section 27 of the 1927 Act do not state what happens if the letter is returned by the Post Office undelivered. In the *Beanby* case, it was argued that if there was evidence that a recorded delivery letter to the tenant containing a section 25 notice had been returned undelivered, the court might be prevented from applying the irrebuttable assumption of delivery because it might be contrary to the rights of the tenant under articles 1 and 6 of the First Protocol of the European Convention on Human Rights. By section 3 of the Human Rights Act 1998, so far as possible to do so, legislation must be read and given effect in a way which is compatible with the "Convention rights" which that Act defines so as to include article 1 of the First Protocol. In that case, Neuberger J left this question unanswered.

Section 23 of the 1927 Act specifically allows for notices to the landlord to be served on the landlord's agent if authorised to receive them (see *Sector Properties Ltd* v *Meah* (1974) 231 EG 357; *Stylo Shoes Ltd* v *Prices Tailors Ltd* [1960] Ch 396). Notwithstanding that section 23 mentions service on an agent only in relation to notices to a landlord, service on a tenant's authorised agent is also effective (*Galinski* v *McHugh* [1989] 1 EGLR 109).

Where the landlord or the tenant comprises two or more persons, ideally they should all sign the relevant notice; if only one signs, it should be stated that he does so on behalf of and with the authority of the others (*Hackney African Organisation* v *Hackney London Borough Council* [1998] EGCS 139, CA).

Service by recorded delivery is in most cases the best method of service. Where physical delivery is preferred, and there are several tenants in a building, the notice should be fixed to the doors of their individual premises and not the main door of the building. It is undesirable to push a notice under a door, as it could go under a carpet or mat and arguments might ensue as to whether it was served. The person serving the notice should record the time and date and manner of service.

Casenotes

Stylo Shoes v *Price Tailors Ltd* [1960] Ch 396: The tenants moved their registered office and main place of business from Huddersfield to Leeds, notified the landlord, and had their mail redirected. The landlord sent a section 25 notice by ordinary post to the Huddersfield address. The Post Office redirected it and it was delivered to the tenants in Leeds. Held: (1) It did reach the tenants, so that was good service. (2) "Place of abode" is equivalent to place of business.

Galinski v *McHugh* [1989] 1 EGLR 109, CA: The landlord's solicitors prepared a notice (under section 4 of the 1954 Act) addressed to the tenant and sent it to the tenant's solicitors, who had told them that they had full authority to act for the tenant and to accept service of the notice. The solicitors did not tell the tenant that they had received the notice until after the time-limit expired for him to claim enfranchisement. Held: There had been valid service on the tenant at the date of delivery of the notice to his solicitors.

Blunden v *Frogmore Investments Ltd* [2002] 2 EGLR 29, CA: The tenant vacated the premises following extensive damage by a bomb. The lease gave the landlord the right to give notice of termination in these circumstances, and stated that notices could be served in accordance with section 196 of the Law of Property Act 1925 or left, addressed to the tenant, at the premises. The landlord served that break notice, together with a section 25 notice, both in accordance with section 196, by recorded delivery post to the last known home address of the tenant and also by fixing the notices on the vacated premises. The tenant was away from home at the time, and, on his return found a communication from the post office that it held a recorded delivery item. However, by then the recorded delivery items had been returned to the landlord's solicitor as undelivered. Held: Service of the break notice was deemed to have taken place by virtue of the express terms of the lease and section 196 of the 1925 Act. Furthermore, the section 25 notice was deemed duly served since posting complied with section 23(1) of the 1927 Act, as incorporated into the 1954 Act, despite the non-delivery.

5.9 Wrong forms and other errors

Where a notice is to be given in a prescribed form, it is important to use the latest version of the correct form. In regulation 2(2) of the 2004 Notices Regulations, it is provided that the notices are to be in the prescribed form set out in schedule 2 to the regulations, "or a form substantially to the same effect". Provisions similar to this appeared in the previous regulations. It means that minor discrepancies from the prescribed form will not matter if a reasonable addressee would not be misled, but any significant error or omission, even in the notes to the form, may invalidate the form.

Errors in completing section 25 notices are discussed separately in Chapter 6 .

If a wrong form of notice is used, or a correct form is incorrectly completed, the recipient should inform the sender that he considers the notice to be invalid. If instead he acts as if he accepts the validity of the notice, and the sender is lulled into believing that it has been accepted and acts in reliance on that, the recipient may be barred from subsequently contending that the notice was invalid. This is because the legal doctrines of estoppel and waiver can be used by the sender against the addressee in such circumstances (*Kammins Ballrooms Co Ltd* v *Zenith Investments (Torquay) Ltd* [1971] AC 850, HL; *Bristol Cars Ltd* v *RKH Hotels* (1979) 251 EG 1279, CA; *Mayhew* v *John Lyon School* [1996] NPC 176, CA).

Casenote

Montgomery v *Sabella Ltd* [1997] EGCS 169, CA: The landlord served a purported section 25 notice, which should have been in a form prescribed in the 1989 Regulations. It omitted the prominent box containing the guidance advising the tenant to act quickly and also omitted five of the notes from the prescribed form. Held: This was not "substantially to the like effect" as the prescribed form and was invalid.

Landlord's Section 25 Notices

6.1 Persons giving and receiving the section 25 notice

One method of terminating a tenancy protected by the Act is for the landlord to serve a section 25 notice. This notice must be given to the tenant, and if appropriate a section 25 notice should also be given to any subtenant, by the competent landlord, who may or may not be the immediate landlord (see 5.1 and 6.5).

Where two or more persons comprise the landlord, it may be that the section 25 notice can validly be given by just one of them (*Leckhampton Dairies Ltd* v *Artus Whitfield Ltd* (1986) 130 Sol J 225), although there is a new specific provision where the reversion is divided between different landlords, discussed below.

If the tenant is an individual who has been made bankrupt and his tenancy has vested by operation of law in his trustee in bankruptcy, the section 25 notice must be addressed to and served on the trustee in bankruptcy (*Gatwick Investments Ltd* v *Radivojevic* [1978] CLY 1768).

Where the premises are occupied by a partnership, as discussed at 2.9, the landlord may, if he wishes, serve a section 25 notice just on the "business tenants" (those partners holding the tenancy who are still in the partnership), omitting those partners who are among the named tenants but who have left the partnership (section 41A(4)). This provision is useful where the landlord cannot readily locate those partners named in the lease who are no longer members of the partnership. The right of those partners who are the "business tenants" to apply for a new tenancy is discussed at 10.5.

A section 25 notice cannot be given to a tenant who has already given the landlord a valid section 26 request for a new tenancy (section 26(4)). However, receiving a section 26 request from the immediate tenant does not prevent the landlord from giving a section 25 notice to a subtenant of that tenant where the subtenant has not yet given a section 26 request.

Where the landlord's reversion has been severed after the grant of lease and the premises divided between two or more different ownerships, the tenant still has just one tenancy of the whole premises comprised in his lease even though he now has more than one landlord. By section 44(1A), introduced by the RRO, where different persons own the relevant reversionary interest to a tenancy, the landlord will be "all those persons collectively". While the view has been expressed that this might allow those persons to give, simultaneously, individual section 25 notices to the tenant, the authors' view is that they should collectively give one section 25 notice. In any event, the substance of the section 25 notice must presumably be the same — either a new tenancy is collectively opposed or terms for a new tenancy are collectively proposed. The Act does not make provision for the situation where one landlord wishes to oppose a renewal and the other does not, or where they disagree over the rent or other terms to be sought for the new lease — that is left for the landlords to sort out between themselves. If the landlords cannot agree, they will be unable to serve a section 25 notice (or a counter-notice to a section 26 request — see Chapter 7).

It is desirable for the landlord to make an early start at least a year before the expiry or break date of the lease. If the landlord delays the service of a section 25 notice, this gives the tenant an opportunity to give the landlord a request under section 26 (see Chapter 7) with the effect of extending the tenancy beyond the contractual expiry date, commonly known as a "pre-emptive strike".

6.2 Premises to be specified

Normally the section 25 notice must relate to the entire premises comprised in the tenant's tenancy and not just part of the premises (*Kaiser Engineers & Constructors Inc* v *ER Squibb & Sons Ltd* (1971) 218 EG 1731, ChD). A possible exception to this rule is where a lease contains a break clause applicable to part only of the premises (see 6.6). The extent of the "holding" is irrelevant to the requirement for the section 25 notice to apply to the whole premises. A single section 25 notice must normally be served on the tenant covering the entire premises let to him (*M&P Enterprises* v *Norfolk Square Hotels* [1994] 1 EGLR 129), but see the discussion about divided reversions at 6.1.

The description of the premises to be inserted into the notice must make it reasonably clear that it is the entire premises comprised in the tenancy. If that is clear, the fact that the wording does not follow the description of the premises in the lease does not matter provided that a reasonable tenant would not be misled (*Safeway Stores Ltd* v *Morris* [1980] 1 EGLR 59; see also *Germax Securities* v *Spiegal* [1979] 1 EGLR 84). If however the description of the premises in the notice is so different so as to give the impression that parts are not included in the notice, the notice will be invalid (*Herongrove Ltd* v *Wates City of London Properties Ltd* [1988] 1 EGLR 82, ChD).

Where the tenant holds two tenancies from the same landlord and those tenancies can be ended on the same date, it has been held that they can be terminated together by one section 25 notice, although this is not good practice and is not recommended (*Tropis Shipping Co* v *Ibex Property Corporation* [1967] EGD 433; *Jones* v *Jenkins* [1986] 1 EGLR 113).

6.3 Form of section 25 notice

The Act specifies what the notice must set out and requires the notice to be given in the appropriate form prescribed by regulations or in substantially similar form. The current forms are prescribed by the 2004 Notices Regulations . There are in effect two categories of forms, one for use where the landlord is willing to grant a new tenancy and one for use where the landlord wants to oppose the grant of a new tenancy. Within each category there are different prescribed forms for use in standard situations and in a variety of special situations. For example, there are special versions for use in relation to the tenancy of a house (see 6.10).

Where the landlord is a public body and wishes to take the benefit of certain special provisions of the Act (see 6.9) there are special forms of section 25 notice that must be used for that purpose. The appropriate form numbers from the Regulations are mentioned below in relation to the various circumstances applicable to public bodies.

If the landlord wants to avoid the risk that the tenant may serve a section 26 request which has the effect of extending the present tenancy (see Chapter 7), the section 25 notice should be given at the earliest opportunity, and not left to just six months before the term expires. However, there are cases where the landlord should consider taking that risk and delay serving his section 25 notice, for example where he will soon be able to oppose a new tenancy but is not yet able to do so.

6.4 Termination date of section 25 notice

Each version of the section 25 notice has a space into which the landlord must insert an expiry date for the current tenancy. That expiry date must be:

(a) not less than six months after the date of giving the notice and
(b) not more than 12 months after the date of giving the notice (except that where the terms of the tenancy provide that it can only be terminated by a notice to quit of a minimum period which is more than six months' duration, a period six months longer than that minimum notice period is substituted for that maximum period of 12 months) and
(c) in the case of a fixed-term tenancy, not earlier than the expiry date of the fixed term or
(d) in the case of a tenancy terminable by notice, not earlier than the earliest date on which a notice to quit given by the landlord would expire if given at the same time as the section 25 notice.

The method of measuring exactly the six and 12 month periods was considered at 5.7 and those rules must be followed here. The eventual postponement of the termination date by subsequent procedures under the Act will be explained in Chapter 10.

There is no requirement for the termination date to be a quarter day, rent day or term anniversary date. It can be the last day of the term of the lease (*Whelton Sinclair* v *Hyland* [1992] 2 EGLR 158) or any subsequent day.

6.5 Notices to subtenants

Where a superior landlord gives a section 25 notice to a subtenant by virtue of being his "competent landlord" (see 5.1), the specified termination date for the subtenancy must not be earlier than the end of the contractual term of the intermediate tenancy. On the other hand, if the intermediate tenancy is continuing after the end of its contractual term, the specified termination date for the subtenancy in the section 25 notice may be earlier than the date when the mesne landlord's intermediate tenancy is due to end under a notice which has been given by the date when the section 25 notice is served (*Lewis* v *MTC (Cars) Ltd* [1975] 1 EGLR 69, CA).

A superior landlord is not obliged to obtain the mesne landlord's consent before serving a section 25 notice on a subtenant, and the notice will not be invalid if he fails to obtain that consent (sixth schedule, para 3(1)). However, it is good practice to apply for that consent, which the mesne landlord cannot unreasonably withhold but which he may grant subject to reasonable conditions, such as for the payment of compensation for loss arising (sixth schedule, para 4).

For example, a subtenancy can be ended by the superior landlord giving the subtenant a section 25 notice expiring in, say, six months time where an intermediate tenancy which is continuing after its contractual term will end after that date under, say, a nine months notice that has just been given.

However, if the superior landlord serves the section 25 notice on the subtenant without the consent of the mesne landlord, or if the mesne landlord so requires as a condition of giving that consent, the superior landlord will be liable to compensate the mesne landlord for the loss of rental income from the subtenant for the period between the end of the subtenancy and the end of the intermediate tenancy.

Further issues relating to subtenants are considered in Chapter 9.

6.6 Relationship with landlords' break notices

Where the contractual term held by a protected tenant is being ended by the landlord serving notice under a provision of the tenancy allowing him to end the contractual term before its expiry date, it is necessary for the landlord also to give the tenant a section 25 notice, as was mentioned in Chapter 4 (*Morrison Holdings Ltd* v *Manders Property (Wolverhampton) Ltd* [1976] 1 EGLR 70). This might apply where the lease contains a landlord's break clause exercisable, say, for redevelopment or following substantial damage. The contractual break notice should be served at the same time as, or immediately before, the section 25 notice. The termination date in the section 25 notice can be the day on which the contractual break notice expires or may be later.

If the landlord serves only a section 25 notice, that notice may be treated as operating the break clause if the contractual break notice does not have to be in any particular form and the notice period under the break clause is consistent with the period of the section 25 notice (*Scholl Manufacturing Co Ltd* v *Clifton (Slim-Line) Ltd* [1967] Ch 41, CA). This equally applies to the termination by the landlord of a periodic tenancy. It is also possible for the landlord to serve a section 25 notice with a covering letter stating that the notice is intended also to be a contractual break notice.

A section 25 notice can be served only in relation to the whole of the premises comprised in a tenancy. This means that normally the termination procedures of the Act cannot be used by a landlord where a break clause allows him to terminate the tenancy as to part only of the let premises (*Southport Old Links Ltd* v *Naylor* [1985] 1 EGLR 66). In that situation, while presumably the giving of the break notice will be effective to end the contractual term as to the part concerned, a s.25 notice will be ineffective and the tenancy of that part will continue under the Act, alongside the contractual term of the remainder, so long as the tenant has protection

In exceptional cases, however, it may be possible to interpret a lease which contains a break clause as to part of the premises as if it created two separate tenancies, one of the terminable part and the other of the remainder. In that unusual situation a section 25 notice relating to just the terminable part of the premises will be valid (*Moss* v *Mobil Oil Co Ltd* (1988) 1 EGLR 71, CA). But in most cases it will not be possible to interpret the lease as creating two separate lettings.

Casenotes

Morrison Holdings Ltd v *Manders Property (Wolverhampton) Ltd* [1976] 1 EGLR 70: Premises were substantially damaged by fire. The tenant urged the landlord to reinstate so that he could resume business in the premises. The landlord served a break notice on the tenant under a clause in the lease which entitled the landlord to demolish the premises after fire damage instead of rebuilding. Held: The landlord would have to go through the section 25 termination procedures under the Act and prove his ground of opposition to a new tenancy in order to get possession. Merely serving a contractual break notice did not end the tenancy.

Southport Old Links Ltd v *Naylor* [1985] 1 EGLR 66: The lease of a golf course entitled the landlord to serve 12 months' notice to recover possession of the whole or any part he might select of the "playing area" (which was a specified part of the golf club). The landlord gave a break notice as to the part he required, and also a section 25 notice relating to it. Held: It was impossible to interpret the lease as creating two separate tenancies, so the purported section 25 notice was void.

Moss v Mobil Oil Co Ltd [1988] 1 EGLR 71, CA: Two separate service stations were comprised in a lease at different rents and the lease contained a provision that it was to be read as if two separate leases had been granted. The landlord served a section 25 notice as to one of the service stations. Held: The section 25 notice as to the specified part of the premises was valid.

6.7 Landlord willing to grant a new tenancy

Where the landlord does not intend to oppose the grant of a new tenancy, the form of section 25 notice to be used will be form 1 in general cases. Special forms are to be used by public body landlords in the special situations mentioned at 6.9.

The form of section 25 notice must be completed by the landlord with the names and addresses of the tenant and the landlord, the address or description of the property and the proposed termination date for the tenancy (see above). In addition, in the schedule to the form the landlord must insert or attach his proposals for three aspects of the new tenancy, the first being the property to be comprised in it (being all or part of the premises comprised in the current tenancy), the second being the rent to be payable under it, and the third being its other terms (section 25(8)). Omitting any of these proposals will invalidate the notice. It is therefore advisable for the landlord's surveyor to inspect the premises before the service of the notice and to carry out all necessary research so as to ascertain how much of the premises are occupied by the tenant (the "holding") and to form a fair opinion of the open market rental value and appropriate lease duration. The landlord should then be advised as to these matters. An appropriate asking rent can and should leave room for negotiation.

It would appear to be contrary to the policy of the Act and the Notices Regulations for the landlord to put "market rent" or some other phrase in relation to the new rent, instead of a sum of money. The prescribed notes to the notice form state that the proposals are "suggestions as a basis of negotiation", and it hardly assists the tenant to commence negotiations if the landlord keeps his proposals vague. Since the prescribed notes make it clear that the proposals are "merely" suggestions, and since the landlord has to insert them in order to commence the termination procedures under the Act, the proposals do not constitute a legally binding offer to the tenant and the landlord will not be prevented from changing them. However, if the matter goes to court and the landlord changes his proposals very substantially, the court might expect some explanation for the changes.

The prescribed notes to the form advise the tenant that the landlord's proposals do not have to be accepted and that the court will decide the terms of the new tenancy if they cannot be agreed. They also warn the tenant of the time-limits for the court application (see Chapter 10) and advise the tenant to take professional advice.

6.8 Landlord opposing a new tenancy

Where the landlord intends to oppose the grant of a new tenancy, he must use the form 2 version of the section 25 notice or, where the premises are a house and the tenant may have a right of enfranchisement under the Leasehold Reform Act 1967 (see the discussion at 6.10), the version in form 7. Special forms are to be used by public body landlords in certain situations (see 6.9).

The notice must state that the landlord intends to oppose any application by the tenant for a new tenancy, and must specify one or more of the statutory grounds set out in section 30 (see Chapter 8).

In the current prescribed forms of section 25 notice, the ground or grounds are to be specified by the landlord only by their letters (a) to (g), but the notes on the reverse of the form explain to the tenant the substance of each ground.

6.9 Public body landlords and public interest cases

Special provisions as to section 25 and section 26 notices apply where the landlord or superior landlord falls within one of the following categories:

(a) Government Departments, not including the Crown Estate Commissioners
(b) local authorities
(c) statutory undertakers
(d) development corporations
(e) health authorities or special health authorities
(f) the National Trust
(g) the Secretary of State or the Urban Regeneration Agency (which took over the assets of the former English Industrial Estates Corporation) in relation to premises in a development area or an intermediate area within section 1 of the Industrial Development Act 1982
(h) the Welsh Development Agency.

Where the landlord's or superior landlord's interest belongs to, or is held for the purposes of, a body falling within category (a) to (e) above, the minister or board in charge of any Government Department can issue a certificate under section 57(1) certifying that it is "requisite" for the purposes of the body to whom the tenancy belongs or for whom it is held (or, in the case of bodies in category (e), for the purposes of the National Health Service Act 1977) that the use or occupation of the property, or part of it, shall be changed by a specified date. The word "requisite" means reasonably necessary and the issue of the certificate can be challenged by the judicial review procedure (*R* v *Secretary of State for the Environment, ex parte Powis* [1981] 1 EGLR 58).

A prior notification and consultation procedure is laid down in section 57(2). The competent landlord or a superior landlord must give the tenant a notice stating that the question of giving the certificate is under consideration by the particular minister or board and that written representations from the tenant given within 21 days will be considered before deciding the question.

If a copy of such a certificate is contained in a section 25 notice which specifies a termination date for the tenancy not earlier than the date mentioned in the certificate, the provisions of section 25(6) (which require section 25 notices to state whether the landlord is opposed to the grant of a new tenancy) will not apply to the section 25 notice and the tenant has no right to apply to the court for a new tenancy (section 57(9)(a)). The form of section 25 notice to use in this case is form 8 or, where the premises are a house (see 6.10), form 13.

If a certificate is issued which specifies a date for changing the use or occupation of the premises which will fall after the termination date that the landlord is inserting in his section 25 notice, the section 25 notice will have to state in the usual way whether the landlord is opposed to a new tenancy and if so on what ground and if not then what terms he proposes for a new tenancy (as required by section 25(6) to (8)), and the tenant will have the right to apply to the court for a new tenancy. However, in that situation the court cannot order a new tenancy to be granted for a term expiring later than the date specified in the certificate as the date for changing the use or occupation of the premises, nor will the new tenancy have the protection of the Act (section 57(3)(b)). The forms of section 25 notice to use in this case are form 9 where the landlord opposes the grant of such a new tenancy or form 10 where he does not oppose it. Where the premises are a house (see 6.10), the form to be used is form 14 which is to have the appropriate insertions and deletions indicating where the landlord opposes or does not oppose the grant of the new tenancy.

If the tenant gives the landlord a section 26 notice requesting a new tenancy at a time when a certificate has been issued under section 57(1), the landlord can, within two months of being given the section 26 notice, give the tenant a notice that the certificate has been given, with a copy of the certificate, in which event the effect on the tenant's right to seek a new tenancy is the same as mentioned above, depending on whether the date specified in the certificate will fall before or after the termination date arising under the section 26 request (section 57(4)(a)).

In cases where the minister or board has started to consider issuing a certificate under section 57(1) but has not yet concluded the matter, it is necessary to prevent the tenant from obtaining a renewal of his tenancy until the decision has been made. Accordingly if, at the time when the tenant gives the landlord a section 26 notice, no certificate has yet been issued but a notice has been given to the tenant under section 57(2) to the effect that consideration is being given to the issue of a certificate, or such a section 57(2) notice is given to the tenant within two months after he gives the section 26 notice, the section 26 notice will be of no effect, but the tenant can serve a new section 26 notice after the minister or board has determined whether to give the certificate.

Alternatively, the minister or board can issue a certificate under section 57(5) that, if the landlord makes an application in that behalf, a court determining the terms of any new tenancy must determine that the new tenancy is determinable by six month's notice to quit given by the landlord. The same consultation procedure as in section 57(2) applies, and if a section 57(2) notice has been given the court must defer determining any pending application for a new tenancy until the minister or board has determined whether to give the section 57(5) certificate. This procedure will be appropriate for cases where the property is likely, at some uncertain future date, to be required for the specified purposes.

Where the landlord's or superior landlord's interest belongs to the National Trust, the provisions of section 57(2) and (4) mentioned above apply, but in this case the certificate given by the minister or board must certify that it is requisite that the use or occupation of the property shall be changed for the purpose of securing that the property will, as from a specified date, be used or occupied "in a manner better suited to the nature thereof" (section 57(7)).

As a further alternative to the section 57(1) procedure, where the landlord's interest belongs to or is held for the purposes of a Government Department, the minister or board in charge of any Government Department can issue a certificate under section 58(1) that for reasons of national security it is necessary that the use or occupation of the property shall be discontinued or changed. In that event a modified section 25 notice can be given, and the tenant loses his right to apply (whether following a section 25 or section 26 notice) for a new tenancy, under provisions similar to those in section 57(3)(a) mentioned above (section 58(1)). The form of section 25 notice to use in this case is form 11.

The provisions of section 58(1) described above also apply where the landlord is the Secretary of State or the Urban Development Agency and the property is in a development area or an intermediate area (designated by order made, or having effect as if made, under section 1 of the Industrial Development Act 1982) and the Secretary of State certifies that it is necessary or expedient for achieving the purpose mentioned in section 2(1) of the Local Employment Act 1972 that the use or occupation of the property should be changed (section 60). The appropriate form of section 25 notice is form 12 or, where the premises are a house (see 6.10), form 15.

When the landlord is the Welsh Development Agency and the Secretary of State certifies under section 60A that it is necessary or expedient, for the purpose of providing employment appropriate to the needs of the area, that the use or occupation of the property should be changed, the provisions of section 58(1) described above will apply (section 60A). In those cases the appropriate form of section 25 notice is form 16 or, where the premises are a house (see 6.10), form 17. Note the provision of Welsh forms mentioned at 1.2.

In addition, section 58(2) provides that nothing in the Act shall invalidate an agreement with the tenant (which can be one of the terms of a tenancy) that the tenancy can be terminated by notice to quit given by the landlord in the event of the minister or board issuing a certificate under section 58(1), and if the tenant is given such a notice to quit containing a copy of such a certificate, the tenancy ceases to have the protection of the Act. Similarly, where the landlord's interest is held by statutory undertakers, section 58(3) provides that nothing in the Act shall invalidate an agreement with the tenant that the tenancy can be terminated by notice to quit given by the landlord in the event of the Minister or Board in charge of a Government Department certifying that possession of the property, or part of it, is urgently required for carrying out repairs (whether on that property or elsewhere) which are needed for the proper operation of the landlord's undertaking. If the tenant is given a notice to quit containing a copy of such a certificate, the tenancy ceases to have the protection of the Act.

Where the tenant is precluded from obtaining an order for the grant of a new tenancy, or from obtaining an order for a new tenancy to be granted for a term beyond a specified date, because of the issue of a certificate under section 57 or section 58, he will on quitting be entitled to compensation calculated under section 37 of the Act (section 59(1)). The assessment of the compensation under section 37 is discussed in Chapter 15. However, this entitlement does not apply where the certificate was given under section 60A and either the premises became vested in the Welsh Development Agency under section 7 of the Welsh Development Agency Act 1975 (being formerly property of the Welsh Industrial Estates Corporation) or under section 8 of that Act (land held under the Local Employment Act 1972) or the tenant was not tenant of the premises when the Agency acquired its interest in the premises (section 59(1A) and (1B)).

6.10 Houses

Special rules apply where the premises comprised in the business tenancy are a "house" within the meaning of the Leasehold Reform Act 1967.

The Leasehold Reform Act 1967 defines a house as including "any building designed or adapted for living in and reasonably so called" (1967 Act section 2). The question whether a particular building can reasonably be called a "house" may be difficult to answer, especially where business use is apparent. In *Tandon* v *Trustees of Spurgeons Homes* [1982] 2 EGLR 73, the House of Lords was divided on the question whether the premises let to the tenant (which in that case had been built and remained as a shop on the ground floor and living accommodation above) could reasonably be called a "house", and it could only reach a majority, not unanimous, decision on the point. This sometimes difficult question has been the subject of a large number of decided cases, which are outside the scope of this book.

A section 25 notice cannot be validly served on a tenant of a house during the subsistence of a claim made by the tenant under the 1967 Act to acquire the freehold or an extended lease of the house (1967 Act, schedule 3, para 2(2)).

If the premises can reasonably be called a house within the meaning of the 1967 Act and the tenant has yet not made a claim to enfranchise under the 1967 Act, the landlord who serves a section 25 notice on the tenant opposing a renewal must use the version of the form of section 25 notice which includes advice to the tenant of his rights under the 1967 Act (1967 Act, schedule 3, para 10). The reason is that a non-renewal will extinguish the tenant's right to enfranchise. Under the 2004 Notices Regulations, the form to be used by ordinary landlords is form 7. Corresponding special versions of the forms are to be used by public body landlords (see 6.9). If a landlord intends to oppose a renewal and it is not certain whether or not the premises are a "house" within the 1967 Act, and he wants the ability to argue that they are not a house, he should serve his section 25 notice on form 7 accompanied by a covering letter stating that in his view the premises are not a house within the meaning of the 1967 Act and that he is serving the section 25 notice using form 7 just as a precaution in case he is wrong but without prejudice to his contention that the premises are not a house.

In these cases, the tenant is given two months from the service of the section 25 notice in which to serve a notice under the 1967 Act claiming the freehold or an extended lease (1967 Act, schedule 3, para 2(1)). If the tenant serves that claim notice within those two months, the operation of the section 25 notice ceases (1967 Act, schedule 3, para 2(2)). If the claim under the 1967 Act then ends without the tenant acquiring the freehold or an extended lease under that Act, the section 25 notice does not revive but the landlord may serve a fresh one. If the landlord does so within one month of the ending of the tenant's claim under the 1967 Act, the earliest termination date which the landlord can specify in the new section 25 notice will be the same termination date as was specified in the previous section 25 notice or, if later, the date three months (instead of the usual six months) from the giving of the fresh notice (1967 Act, schedule 3, para 2(3)).

The right to purchase the freehold is very valuable, particularly where the lower limits in the Leasehold Reform Act 1967 apply, because the purchase price payable by the tenant for the freehold is, in effect, discounted. In view of this, the tenant's advisers should always check the relevant provisions of the 1967 Act to see whether enfranchisement might be possible, which valuation basis under the Act would apply and whether the landlord has served the correct version of the section 25 notice.

Forms of section 25 notice which advise tenants of rights under the 1967 Act are also prescribed for use where the landlord is a Government department or other public body (see 6.9).

6.11 Amending and withdrawing section 25 notices

Once served, a section 25 notice cannot be withdrawn except in the special circumstances of 1954 Act sixth Schedule, para 6 using form 6 (see 9.2).

A landlord cannot amend a section 25 notice in order to add grounds of opposition which he claims to have but which he did not specify in the notice, although he may be able to put forward such matters at the court hearing in order to influence the terms of renewal that the court will order. In *Amika Motors Ltd* v *Colebrook Holdings Ltd* [1981] 2 EGLR 62, the landlord neglected to oppose renewal on ground (f) but used the existence of a planning permission for redevelopment to persuade the court to insert a redevelopment break clause into the new lease.

A landlord can withdraw grounds of opposition, but even if he withdraws all grounds which would entitle the tenant to compensation under section 37 (see Chapter 14), this will not avoid liability to pay compensation if the tenant decides to vacate.

6.12 Section 25 notices given "without prejudice"

Normally a section 25 notice should be served without qualifying it in any way. It should certainly not have any extraneous words added to the notice itself, apart from filling in the required insertions in the blank spaces.

Nevertheless, where the landlord is uncertain whether a tenant or occupier is protected by the Act (perhaps he is not sure if the tenant is in occupation or is carrying on a business, or if he may be a mere licensee) it is possible to serve a section 25 notice accompanied by a covering letter stating that the landlord believes that the tenant is unprotected but that the landlord is serving the notice without prejudice to the landlord's contention on that point and merely as a precaution in the event that the landlord may be found to be wrong.

6.13 Errors in section 25 notices

The use of correct prescribed forms under the Act was considered at 5.9, and the range of prescribed section 25 notice forms is described earlier in this chapter. One should ideally complete the correct form of section 25 notice as accurately as possible. However, committing minor errors in completing the form may not render it void, provided that it would not mislead a reasonable tenant. This is consistent with the general approach of the law to errors in notices established by the leading case of *Mannai Investment Co Ltd* v *Eagle Star Life Assurance Co Ltd* [1997] 1 EGLR 57, HL; although that case related to a contractual notice, the same rule applies to statutory notices (*York* v *Casey* [1998] 2 EGLR 25, CA).

If the landlord suspects that a section 25 notice that he has served was invalid, he can serve a fresh one accompanied by a covering letter stating that the fresh notice is given without prejudice to his contention that his earlier notice was valid and also stating that the fresh notice is being given as a precaution in case that contention is proved wrong (*Draper* v *Smith*, below).

As for the tenant's response to a notice which he suspects may be invalid, see 5.9 and 6.14.

Casenotes

Morrow v *Nadeem* [1986] 2 EGLR 73: The landlord company's name was omitted from the body of the s.25 notice and the notice was signed by "Solicitors and agents for" someone who was not in fact the landlord but was the individual who owned the landlord company. Held: The notice was invalid.

Yamaha-Kemble Music (UK) Ltd v *Arc Properties* [1990] 1 EGLR 261: After the original landlord's interest was transferred to a subsidiary company, a s.25 notice was served which by mistake named the original landlord as the landlord, rather than the subsidiary which had become the landlord. Held: The notice was invalid.

Pearson v *Alyo* [1990] 1 EGLR 114, CA: Mr and Mrs Alyo were joint landlords but a section 25 notice was served naming only Mr Alyo as landlord. Held: The notice was invalid even though the tenant was not actually prejudiced, since objectively viewed the notice was capable of prejudicing a reasonable tenant.

Smith v *Draper* [1990] 2 EGLR 69, CA: The reversion was owned by three joint landlords but a s.25 notice only named two of them. Held: The notice was invalid.

Stidolph v *American School in London* (1969) 211 EG 925: An unsigned notice was sent with a signed covering letter. Held: The signing of the covering letter was sufficient to render the notice valid.

Germax Securities v *Spiegal* [1979] 1 EGLR 84: A notice containing an invalid termination date was accompanied by a covering letter referring to the correct date. Held: The notice was valid as if it contained the date mentioned in the covering letter.

Carradine Properties Ltd v *Aslam* [1976] 2 All ER 573: A section 25 notice served towards the end of 1974 specified the year of the termination date as 1973 instead of the obviously intended 1975. Held: The notice was valid as if it had specified 1975.

Bridgers v *Stanford* [1991] EGCS 50, CA: A lease comprising only the ground floor and first floors of a building at 70 High Street, Epsom, was granted to Bridgers, a firm of estate agents. They were later taken over by Hamptons and traded as Bridgers Hamptons. The landlord served a section 25 notice naming the tenant as "Bridgers" and describing the premises as "70 High Street, Epsom". Held: The notice was valid.

Herongrove Ltd v *Wates City of London Properties plc* [1988] 1 EGLR 82: A lease comprised ninth floor offices, lower ground floor and basement storage areas, and three parking spaces. The landlord served a section 25 notice referring only to the ninth floor offices. Held: The notice was invalid.

Barclays Bank plc v *Bee* [2001] 3 EGLR 41, CA: The landlord told the tenant that he intended to redevelop the property in two or three years. A few months later the landlord's solicitors served two conflicting section 25 notices at the same time. One, which was valid on the face of it, did not oppose a renewal and the other purported to oppose renewal but omitted to set out any ground of opposition, and was invalid. When this was pointed out, they claimed that both notices were invalid and served a fresh notice opposing renewal. The tenant claimed that it could rely on the landlord's non-opposing notice. Held: The original notices were void and the fresh notice was valid. Although the non-opposing notice might alone have been valid, a reasonable recipient of the letter enclosing the two conflicting notices would have realised that a mistake had been made and should not be entitled to choose which notice to treat as valid.

6.14 Tenant's response to a section 25 notice

In order to obtain a new tenancy following receiving a section 25 notice, the tenant must, before the termination date specified in the notice, either actually be granted the renewal tenancy, or agree terms of renewal with the landlord in a legally binding way, or apply to the court, or obtain the landlord's written agreement to extend the time for applying to the court. The last two processes will be described in Chapter 10.

If the tenant does not want a new tenancy, and the termination date specified by the landlord in the section 25 notice is later than the expiry date of the contractual term , the tenant may be able to vacate and terminate the tenancy sooner, as discussed in Chapter 4.

If the tenant claims that the landlord's section 25 notice is invalid for some reason, the tenant should challenge it without delay (see 5.19) and he should consider whether it would be beneficial to serve his own section 26 request on the landlord (see Chapter 7). In the event of dispute, the validity of a notice can be referred to the court for decision

In cases where the section 25 notice was served before 1 June 2004, the unamended version of the Act applies and, in order to retain protection, the tenant would have had to serve a counter-notice on the landlord within two months of the giving of the section 25 notice, stating that the tenant was not willing to give up possession on the termination date (old section 25(5), old section 29(2)); he would also have had to apply to the court for a new tenancy between two and four months after the giving of the section 25 notice (old section 29(3)). The requirement for the tenant to serve a counter-notice does not apply to section 25 notices served after 1 June 2004 and the time-limit for the court application is longer, as mentioned above.

Tenant's Section 26 Requests

7

7.1 Tenants entitled to request new tenancies

A section 26 request for a new tenancy may be given only by tenants whose tenancy falls within certain criteria. By section 26(1), the request can be given only by a tenant

(a) whose current tenancy is for a term certain exceeding one year or
(b) whose current tenancy was granted for a term certain exceeding one year and thereafter from year to year or
(c) who is occupying under a section 24 continuation tenancy after the expiry of such a tenancy as is mentioned in either (a) or (b) above.

Under these provisions, a tenant holding only a periodic tenancy cannot give a section 26 request, and a tenant who ends a periodic tenancy by giving a notice to quit loses his security under the Act (see 4.5). The combination of these rules, although perfectly consistent with each other (and with the decision in *Garston* v *Scottish Widows Fund* [1998] 2 EGLR 73, CA, cited at 4.5), has important consequences for such a tenant, especially in a falling market. Unlike a tenant who is holding over under a continuation tenancy following the expiry of a fixed term lease, a periodic tenant who has not previously held under a fixed term tenancy cannot take steps to end his tenancy and still have the right under the Act to seek a new tenancy at market rent.

A tenant who is otherwise entitled to give a section 26 request cannot do so after he has been given a valid section 25 notice by his competent landlord (section 26(4)).

Where the property comprised in the tenancy is a house (see 2.15 and 6.10), the tenant cannot validly serve a section 26 notice during the currency of a claim he has made under the Leasehold Reform Act 1967 to acquire the freehold or an extended lease of the house (1967 Act, schedule 3, para 1(2)).

7.2 Making a section 26 request

As in the case of section 25 notices, section 26(3) the Act specifies what the section 26 request must set out and provides for the form of request to be prescribed by regulations. It is currently form 3 prescribed in the 2004 Notices Regulations.

The form must set out the tenant's proposals for the property to be comprised in the new tenancy (being the whole or part of the property comprised in the tenant's current tenancy), the rent to be payable, and the other terms of the new tenancy, which would include its duration (section 26(3)).

The request must be addressed to and given to the competent landlord, who may or may not be the immediate landlord (see Chapter 5).

There is no requirement for the tenant to be genuine in his expressed desire to have a new tenancy. He can serve a section 26 notice for purely tactical purposes and then refrain from following it up (*Sun Life Assurance plc* v *Thales Tracs Ltd* [2001] 2 EGLR 57, CA).

7.3 Commencement date for the new tenancy

Unlike a section 25 notice, which has to specify the date on which the landlord wants the current tenancy to end, the section 26 request must state the date on which the tenant wants his new tenancy to commence. However, the rules about the date are similar.

To comply with section 26(2), the section 26 request must specify a date for commencement of the new tenancy which is:

(a) not less than six months after the date of giving the request
(b) not more than 12 months after the date of giving the request
(c) in the case of a fixed-term tenancy which has not yet expired, not earlier than the expiry date of the fixed term
(d) in the case of a continuation after the expiry of a fixed-term tenancy, not earlier than the earliest date on which a notice to quit given by the tenant would expire.

As with a section 25 notice, the termination date need not be a quarter day, rent day or term anniversary. The method of measuring time-limits is discussed at 5.7.

It will be apparent from the above that a tenant can, by suitable timing, extend his existing tenancy by giving a section 26 request which will expire after the contractual expiry date of his lease. Whether he should do so depends upon the circumstances.

The effect of serving a section 26 request is to end the current tenancy on the day before the date specified in the request for the start of the new tenancy. However, if an application to the court is made, this will be extended by section 64 (see 10.23).

7.4 Landlord's counter-notice

If the landlord receives a section 26 notice from the tenant requesting a new tenancy, the landlord has two months in which to serve a counter-notice if he intends to oppose the grant of a new tenancy. The exact measurement of this time limit is considered at 5.7.

The counter-notice must specify the ground(s) under section 30(1) on which the landlord intends to oppose the renewal (section 26(6)). There is no prescribed form of counter-notice.

Failure to give a counter-notice within the prescribed period will lose the landlord the opportunity to oppose the grant of a new tenancy, so the tenant will have an absolute entitlement to be granted a new tenancy, although the terms will still need to be agreed or determined by the court (section 30(1)).

Landlord's Opposition to Renewal

8.1 Overview of the grounds of opposition

There are seven grounds of opposition set out in section 30(1) of the Act. These are set out and analysed in the following sections of this chapter. They are designated (a) to (g) in section 30(1) and briefly are as follows:

(a) tenant's failure to repair
(b) persistent rent arrears
(c) other defaults
(d) offer of suitable alternative accommodation
(e) subletting of part reducing value
(f) demolition or construction
(g) landlord intending to occupy.

If the landlord includes one or more of these grounds in a section 25 notice or section 26(6) counter-notice, in opposition to the grant of a new tenancy, the tenant can put the landlord to proof of the existence of the relevant ground or grounds in court (see Chapter 10).

8.2 Ground (a) — failure to repair

> Where under the current tenancy the tenant has any obligations as respects the repair and maintenance of the holding, that the tenant ought not to be granted a new tenancy in view of the state of repair of the holding, being a state resulting from the tenant's failure to comply with the said obligations.

For this ground to apply, there must be an obligation to repair contained as a term of the present tenancy, and the failure to repair must relate to "the holding"; failure to repair other parts of the premises comprised in the tenancy is irrelevant (damages will be available to the landlord in respect of these parts). Furthermore, the holding must be in disrepair at the date of the hearing; if substantial repairs have been carried out by that date, that can be taken into account by the court (*Hazel* v *Akhtar* [2002] 1 EGLR 45, CA).

The court has a discretion under this ground, since the paragraph uses the words "ought not" to be granted a new tenancy. A renewal is not automatically refused merely because of the tenant's failure to repair, but where that exists the court can also consider other facts relating to the tenant's conduct, in addition to the breach of repairing obligation, when deciding how to exercise its discretion (*Hutchinson* v *Lambeth* [1984] 1 EGLR 75, CA).

If a landlord wishes to pursue this ground, he should prepare his case by generating correspondence with the tenant insisting on the execution of the repairs, and by serving schedules of dilapidations on the tenant, ideally during the year or years preceding service of the section 25 notice. Evidence of proceedings for forfeiture on the ground of disrepair will be useful, even if they had been compromised, or not pursued. The purpose of showing this prior action is to demonstrate the tenant's disregard of his liability under the lease.

If the landlord is in danger of failing to persuade the court to refuse the tenant a new tenancy on ground (a), he might consider asking the court to include in the new tenancy, if ordered, a provision obliging the tenant to carry out specified repairs within a specified period. An agreed schedule of repairs would probably be needed for this purpose.

Casenote

Eichner v *Midland Bank Executor & Trustee Co* [1970] 2 All ER 597: In considering the landlord's opposition under grounds (a) and (c) of section 30(1), the judge took account of a great deal of litigation between the parties over the years and whether it was fair to saddle the landlord with that tenant in those circumstances. On appeal, the Court of Appeal held: The judge is entitled to consider the conduct of the tenant as a whole in relation to all the terms of his tenancy, and is not confined to the grounds stated in the notice.

8.3 Ground (b) — persistent rent arrears

> That the tenant ought not to be granted a new tenancy in view of his persistent delay in paying rent which has become due.

The delays in paying rent must have been persistent. There need be no arrears at the date of the hearing; one looks at the history of the whole duration of the tenancy. However, the landlord will not be able to point to delayed payment of rent during a past period when the landlord (including any previous landlord) had acquiesced in that behaviour and had not pressed for prompt payment (*Hazel* v *Ahktar* [2002] 1 EGLR 45, CA).

As under ground (a), this ground is discretionary involving a judgment as to whether the tenant "ought not" to be granted a new tenancy. The court has a discretion and primarily considers whether future delays are likely to occur, although the tenant's conduct generally can be taken into account (*Eichner* v *Midland Bank Executor & Trustee Co* [1970] 2 All ER 597, CA; *Hutchinson* v *Lambeth* [1984] 1 EGLR 75, CA).

The landlord should ensure that the tenant is chased for prompt payment of rent. The landlord should keep a rent history covering the amounts and dates of all the tenant's payments during the term of the lease including, and in particular, a record of any bouncing cheques.

In appropriate cases the landlord who fails to succeed on ground (b) might nevertheless be able to persuade the court to require that any new tenancy must include a rent guarantee from a suitable guarantor (see *Cairnplace Ltd* v *CBL Ltd* [1984] All ER 315, at 12.4).

Casenotes

Hopcutt v *Carver* (1969) 209 EG 1069: The tenant had constantly been in arrears with the rent over the preceding two years. Held: Ground (b) was proved and a new tenancy was refused.

Hurstfell Ltd v *Leicester Square Properties Ltd* [1988] 2 EGLR 105: All payments of rent between 1984 and the end of 1986 were between four weeks and 19 weeks late. However, rent was paid promptly after the landlord served a section 25 notice early in 1987 opposing renewal on ground (b). Held: The discretion of the court involves considering whether it was likely that further arrears would occur under any new tenancy. (The Court of Appeal felt unable to overrule the trial judge's conclusion that future delays were unlikely, although stated that they themselves might have reached a different conclusion on the available accounting evidence.)

Rawashdeh v *Lane* [1988] 2 EGLR 109: The trial judge had refused a new tenancy, deciding that "persistent" delay had been made out on the evidence that all but the first and last payments of rent under the 21 year lease had been late, four cheques for rent had not been initially honoured, and there had been two set of possession proceedings. On appeal the Court of Appeal held: The judge had not erred in principle nor was his decision perverse.

Hazel v *Akhtar* [2002] 1 EGLR 45, CA: The tenant was persistently late in paying rent but his previous landlord never complained over the first 14 years of the lease. The only complaints about late payment were made in relation to the rent that fell due shortly before the new landlord served a section 25 notice opposing renewal on ground (b). Held: The delays that had been tolerated by the person who was landlord at the time could not be invoked for the purpose of ground (b), so the landlord could not claim that there was "persistent" late payment.

8.4 Ground (c) — other breaches or reasons

> That the tenant ought not to be granted a new tenancy in view of other substantial breaches by him of his obligations under the current tenancy, or for any other reason connected with the tenant's use or management of the holding.

This ground of opposition to renewal involves two alternative, or cumulative, types of default by the tenant, the first being breaches of lease obligations (other than repairing and paying rent) and the second being other reasons connected with the tenant's use or management of the holding.

To fall within the first category, the breach of tenancy obligations must be "substantial". No criteria for this substantiality are set out in the Act but probably it means that the breach must have some substantial adverse effect on the landlord or the value of his reversion. The court can take into account any number of breaches. Although the breaches of covenant will usually be connected with the holding, they apparently need not be, and could relate to other parts of the premises comprised in the existing tenancy, including any sublet parts.

The court can take into account any number of breaches. Although the breaches of covenant will usually be connected with the holding, they apparently need not be, and could relate to other parts of the premises comprised in the existing tenancy, including any sublet parts.

The second category (other reasons connected with use or management) can be invoked even where the tenant is not actually in breach of any obligation under the lease. However, this category of default is limited to matters relating to the "holding", being the parts occupied by the tenant himself, and is confined to aspects of the tenant's use or management of the holding. The tenant's management of any parts of the premises which he sublets is not relevant to this second category of ground (c) since those parts are not within the holding.

As in the cases of grounds (a) and (b), this is a discretionary ground involving a judgment as to whether the tenant "ought not" to be granted a new tenancy. The tenant's conduct generally can be taken into account (see *Eichner* v *Midland Bank Executor & Trustee Co* [1970] 2 All ER 597, CA; *Hutchinson* v *Lambeth* [1984] 1 EGLR 75, CA).

The history of the tenancy will be important when the court considers how to use its discretion. The court will usually want evidence that the landlord was adversely affected by the tenant's breaches, misuse or mismanagement. As with the other discretionary grounds, a history of previous complaints from the landlord to the tenant, with details of any unfulfilled assurances that the tenant may have given to the landlord, will assist the landlord in supporting his case

Casenotes

Beard v *Williams* [1986] 1 EGLR 148: The terms of a weekly tenancy permitted the tenant to use the land for breeding or training greyhounds, but prohibited him from living there. In breach, the tenant brought a decrepit van onto the land and slept in it. The landlord obtained an injunction and the tenant moved the van onto the road just outside the land, claiming it was necessary to live at or adjacent to the land. Although the van was parked illegally, the local authority took no action. The landlord served a section 25 notice opposing a renewal on ground (c), and argued that in view of the tenant's claim that the business required him to live near the premises, no renewal should be granted since there was no lawful residential accommodation available for the tenant nearby. On appeal, the Court of Appeal held: The precarious nature of the tenant's living arrangements could properly be taken into account by the judge in deciding whether a future deterioration of the business was likely to occur, to the prejudice of the landlord. (The case was remitted back to the county court judge to exercise his discretion.)

Turner & Bell v *Searles (Stanford-Le-Hope Ltd)* [1977] 2 EGLR 58, CA: The premises, let on an oral monthly tenancy, were used as a coach depot. The local authority served an enforcement notice since this use was contrary to planning control. The tenant appealed unsuccessfully. The council, however, did not pursue the matter because they used the tenant's coaches for school transport. The landlord served a section 25 notice opposing renewal on ground (c). Held: The scope of the second limb of ground (c) was not limited to matters within the landlord and tenant relationship, and could encompass the lawfulness of the use under planning law. The court was bound to refuse a new tenancy since to order the grant of one would require the parties to enter into an illegal contract which the court could not enforce.

8.5 Ground (d) — offer of suitable alternative accommodation

> That the landlord has offered and is willing to provide or secure the provision of alternative accommodation for the tenant, that the terms on which the alternative accommodation is available are reasonable having regard to the terms of the current tenancy and to all other relevant circumstances, and that the accommodation and the time at which it will be available are suitable for the tenant's requirements (including the requirement to preserve goodwill) having regard to the nature and class of his business and to the situation and extent of, and facilities afforded by, the holding.

The alternative accommodation must be offered to the tenant before, and be available at the date of, the court hearing and it must remain available at the date on which the present tenancy will come to an end under section 64 (see 10.23). At first glance, the wording of ground (d) appears to require the offer to have been made before the section 25 notice, or the counter-notice under section 26(6), is served. However it has been held in an unreported county court case that, because the notice simply sets out the ground on which the landlord would oppose a request by the tenant to the court asking for a new tenancy if the tenant made one, there is no need for the landlord to have offered the accommodation prior to giving the section 25 notice, provided that it was offered on reasonable terms prior to the issues being joined in the court proceedings (*M Chaplin Ltd* v *Regent Capital Holdings Ltd* (1983) Judge Aron Owen, Clerkenwell County court, unreported). The terms offered by the landlord should include the amount that he is willing to pay in respect of the tenant's removal costs.

It has been observed that ground (d) does not make clear the date on which the suitability of the alternative accommodation is to be assessed; in one case the tenant objected to the accommodation proposed and the court allowed the landlord to revise his offer at the hearing to meet the objections (*Mark Stone Car Sales Ltd* v *Howard de Walden Estates Ltd* (1997) 73 P&CR 43, CA).

If the accommodation is found to be suitable but its availability will be delayed, the tenant will be entitled to a new tenancy of his present holding unless section 32(2) applies. This enables the landlord to succeed even if the alternative accommodation will not be available on the termination date of the current tenancy (usually three months and two weeks after the court order — see 10.23), provided that the landlord satisfies the court that the date on which it will be available falls within a year of the termination date as originally specified in the section 25 notice or section 26 request. In many cases, though, the delays inherent in the procedures under the Act will result in the hearing not taking place until after that specified termination date has long passed, so the time extension under section 31(2)(a) may rarely give real benefit to the landlord.

Whether the requirements of paragraph (d) are satisfied in any particular case is a question of fact. The terms on which the accommodation is being offered should include the payment of removal costs and necessary costs of adaptation of the accommodation, and the terms of the new lease as regards rent, repairing obligations etc should correspond with those terms under the existing tenancy. The terms must be reasonable as at the date of the hearing and may be changed by the landlord at the hearing from those that he had previously been offered. An offer of the alternative accommodation on terms to be determined by the court is thought to be a satisfactory offer.

In some cases it may be appropriate for the landlord to offer, as the alternative accommodation, part only of the premises comprised in the existing tenancy (*Lawrence* v *Carter* (1956) 167 EG 222). For example, if the tenant of an entire floor of a building has sublet half of it outside the protection of the Act and the landlord wishes to occupy for his own business the half still occupied by the tenant, he might offer the tenant just the sublet part, possession of which will become available at the end of the term, as the alternative accommodation.

8.6 Ground (e) — subtenancy of part

> Where the current tenancy was created by the subletting of part only of the property comprised in a superior tenancy and the landlord is the owner of an interest in reversion expectant on the termination of that superior tenancy, that the aggregate of the rents reasonably obtainable on separate lettings of the holdings and the remainder of that property would be substantially less than the rent reasonably obtainable on a letting of that property as a whole, that on the termination of the current tenancy the landlord requires possession of the holding for the purpose of letting or otherwise disposing of the said property as a whole, and that in view thereof the tenant ought not to be granted a new tenancy.

This ground applies where the landlord's immediate tenant has sublet part of the premises demised to him, the immediate tenant is not being given a new tenancy, and the subtenant is seeking a new tenancy of the part that was sublet to him.

It has been held, but only at first instance, that this ground can only be used if, at the date of the subtenant's application to the court, the intermediate lease is still subsisting so that the competent landlord "is the owner of an interest in reversion expectant on the termination of that superior tenancy". On that basis this ground cannot be used if the intermediate tenancy ends before that date (*Greaves Organisation Ltd* v *Stanhope Gate Property Co Ltd* (1973) 228 EG 725). In some cases the intermediate tenancy will in fact continue beyond the date of the subtenant's application — for example where the head tenant is still carrying on business in the part he has retained, so that his tenancy is itself continuing under the Act — but this will not always be the case.

The landlord must be able to show that the rent for the premises comprised in the immediate tenant's tenancy can be let for substantially more as a whole than by separate lettings of the sublet part and the remainder. This ground is little used, probably because in many cases the aggregate rental value of the parts will exceed that of the whole, rather than vice versa. The difference in capital value (which might be enhanced by a single letting even at the same rent as that achievable from multiple lettings) is irrelevant; this ground is confined to a difference in rental value, and that the difference has to be "substantial". Again, the court has a discretion whether or not to reject a renewal on this ground.

The landlord must require possession for letting or other disposal "at the termination of the current tenancy", usually three months and two weeks after the case is finally disposed of (see 10.23). However, by section 31(2)(a), the landlord will succeed even if possession for that purpose is not required until a later date, so long as that is still within a year of the termination date originally specified in the section 25 notice or section 26 request. In many cases, though, the delays inherent in the procedures under the Act will result in the hearing not taking place until after the specified termination date has long passed, so the time extension under section 31(2)(a) may rarely give real benefit to the landlord.

This ground is usually only applicable where the subletting of part actually reduces the value of the whole. One example of this is where the premises as a whole would be valued as clear space but, in order to create one or more areas for sublettings, the tenant had erected partitions to create corridors, materially reducing the lettable area. Another example is where premises are shared without any structural division. In some places it is still common to find older tobacconist's and newsagent's shops where the rear area of the shop is sublet to a barber who has a working relationship with the tenant who runs the tobacconist/newsagents. If the tenant left, it would be difficult for the landlord to re-let the entire premises to a new tenant if the barber remained in possession of the rear part.

Casenote

Greaves Organisation Ltd v *Stanhope Gate Property Co Ltd* (1973) 228 EG 725: The landlord let the lower part of its building. The lease was terminated by a break notice. The tenant applied for a new tenancy and the landlord agreed to grant a 12 year lease. The tenant then agreed to grant a sublease of one floor for the length of the proposed new lease less three days, and the subtenant moved in. The new lease was never completed, the tenant giving up possession. The landlord then served a section 25 notice on the subtenant opposing renewal on ground (e). Held: The agreement for the 12 year lease had been binding and entitled the tenant to grant the subtenancy in equity, and this was binding on the landlord, but the landlord would not have succeeded on ground (e) in any event because, first, although he could show that the aggregate rents obtainable would be less, they would not be "substantially" less than the rent for the whole lower part of the building, and, secondly, the superior tenancy had ended prior to the subtenant's court application.

8.7 Ground (f) — demolition and construction

That on the termination of the current tenancy the landlord intends to demolish or reconstruct the premises comprised in the holding or a substantial part of those premises or to carry out substantial work of construction on the holding or part thereof and that he could not reasonably do so without obtaining possession of the holding.

Although this is not the longest-worded ground of opposition to renewal, it comprises many separate elements, leading to a vast number of court cases in which the interpretation of every word of the ground has been argued over.

This ground is supplemented by section 31A:

> (1) Where the landlord opposes an application under section 24(1) of this Act on the ground specified in paragraph (f) of section 30(1) of this Act, or makes an application under section 29(2) of this Act on that ground, the court shall not hold that the landlord could not reasonably carry out the demolition, reconstruction or work of construction intended without obtaining possession of the holding if —
>
> (a) the tenant agrees to the inclusion in the terms of the new tenancy of terms giving the landlord access and other facilities for carrying out the work intended and, given that access and those facilities, the landlord could reasonably carry out the work without obtaining possession of the holding and without interfering to a substantial extent or for a substantial time with the use of the holding for the purposes of the business carried on by the tenant; or
>
> (b) the tenant is willing to accept a tenancy of an economically separable part of the holding and either paragraph (a) of this section is satisfied with respect to that part or possession of the remainder of the holding would be reasonably sufficient to enable the landlord to carry out the intended work.
>
> (2) For the purposes of subsection (1) (b) of this section a part of a holding shall be deemed to be an economically separate part if, and only if, the aggregate of the rents which, after the completion' of the intended work, would be reasonably obtainable on separate lettings of that part and the remainder of the premises affected by or resulting from the work would not be substantially less than the rent which would then be reasonably obtainable on a letting of those premises as a whole.

In *Fisher* v *Taylors Furnishing Stores* [1956] 2 All ER 78, ground (f) was analysed by Morris LJ as requiring a provable intention to carry out works falling within at least one of the following six categories:

(1) demolition of the whole of the premises comprised in the holding
(2) reconstruction of the whole of the premises in the holding
(3) demolition of a substantial part of the premises comprised in the holding
(4) reconstruction of a substantial part of the premises in the holding
(5) carrying out of substantial work of construction on the holding and
(6) carrying out of substantial work of construction on part of the holding.

In general, these categories require that the scheme which the landlord intends to carry out must involve demolition, construction or reconstruction and that these works, viewed as a whole, must be substantial. One looks at the whole of the landlord's proposed work for this purpose rather than taking each element individually (*Bewlay (Tobacconists) Ltd* v *British Bata Shoe Co Ltd* [1958] 3 All ER 652, CA).

Many of the early reported cases on ground (f) contain statements that the work had to affect "structural" or "load-bearing" elements of the property, since the courts identified "construction" and "reconstruction" with the creation or alteration of "structures". It was said that "reconstruction" meant a measure of structural demolition coupled with the construction of a new structure in a different form (*Cook* v *Mott* (1961) 178 EG 633; *Cadle* v *Jacmarch* [1957] 1 All ER 148). Similarly, "work of construction" was said to mean the erection of a new structure, which might include a wall, concrete platform or road (*Botterill* v *Bedfordshire County Council* [1985] 1 EGLR 82). Those interpretations may still be correct where the premises demised by the tenant's existing tenancy include structures and load-bearing elements. Where however the premises demised by the tenancy exclude the structural parts of the building, the approach is different and the removal of enclosing walls, floors and ceilings of the holding can still be considered to involve "demolition" and "reconstruction" of the holding even where the holding is confined to non-structural surfaces, sometimes called the internal "eggshell" (*Pumperninks of Piccadilly Ltd* v *Land Securities plc* [2002] 2 EGLR 147, CA; *City Offices (Regent Street) Ltd* v *Europa Acceptance Group plc* [1990] 1 EGLR 63, CA).

In the first four categories listed by Morris LJ, it appears that the "premises comprised in the holding" does not mean the holding itself but connotes buildings and other structures on the holding. This distinction may be important where the holding is a parcel of land with buildings or other structures on only part of it; it seems that the landlord will satisfy ground (f) if he intends to demolish or reconstruct just those buildings and structures even if he will not carry out any substantial work elsewhere on the holding (*Housleys* v *Bloomer-Holt* [1966] 2 All ER 966, CA).

The relevant works are those that the landlord in fact intends to do, and the landlord is entitled to design his scheme having in mind the need to fall within ground (f). So long as the landlord actually intends to carry out those works, the tenant cannot argue that different works, falling outside ground (f), would be preferable (*Decca Navigator* v *Greater London Council* [1974] 1 WLR 748).

Where the premises comprised in the holding will not be substantially demolished, in order to fall within ground (f) the intended work must involve construction or reconstruction which is "substantial". All the construction work to be carried out on the holding must be viewed as a whole, rather than considering component items individually, in order to see if the work is "substantial". It is permissible to take into account not only the construction work itself but also necessary closely associated work, such as plastering a newly constructed wall. However, works not closely connected with construction are to be ignored (*Romulus Trading Co Ltd* v *Trustees of Henry Smith's Charity Trustees* [1990] 2 EGLR 75, CA).

The "intention" must be that of the landlord, although he may intend to carry out the development with or through others. The landlord may have the necessary intention if he intends to carry out the works himself, or intends to engage a building contractor to do them, or intends to enter into an arrangement with a third party, such as a building lease or building agreement which requires the works to be carried out by that party (*Gilmour Caterers Ltd* v *St Bartholomew's Hospital Governors* [1956] 1 QB 387, CA; *Peter Goddard & Sons Ltd* v *Hounslow London Borough Council* [1992] 1 EGLR 281, CA). A landlord who intends to sell the property after obtaining possession but before development takes place would not normally come within this ground, since the person who will redevelop the property will not have been the tenant's landlord at any time and ground (f) requires the "landlord" to intend to carry out the work. However, ground (f) might possibly apply in such cases if the landlord imposes an obligation on his purchaser to carry out the works for their joint benefit, perhaps under an "overage" arrangement where he participates in the development value of the property. However, in order to succeed on ground (f), it might be safer in such cases for the landlord to follow the cases cited above and enter into an agreement for lease under which the intending lessee is to carry out the development and is to be granted a long lease of the premises, even if the lease is to contain an option for the lessee to purchase the freehold for a nominal sum. On that basis the landlord still has ownership of the premises when the work is carried out and retains a degree of control over the work.

The fact that the landlord may intend to sell the property after the work of demolition and construction does not prevent him successfully opposing renewal on ground (f) (*Turner* v *Wandsworth London Borough Council* [1994] 1 EGLR 34, CA).

As for cases where a change of competent landlord occurs between the giving of the section 25 notice (or counter-notice for a section 26 request) and the court hearing, see 8.9.

Simply showing that redevelopment as desirable or valuable will not be sufficient if is not coupled with evidence that the landlord "intends" to carry it out. There must be a firm and *bona fide* decision to carry out the work if possession is obtained and the "reasonable prospect" of having the ability to carry it out.

The need for a firm and *bona fide* decision to carry out the work, which derives from the use of the word "intends" in ground (f), is based on the classic exposition of the meaning of "intends", which is that the matter must move "from the zone of contemplation — out of the sphere of the tentative, the provisional and the exploratory — into the valley of decision" (Asquith LJ in *Cunliffe* v *Goodman* (1950) 155 EG 202, CA, regarding the landlord's intention to redevelop within section 18 of the Landlord and Tenant Act 1927). This may be a matter of degree. In one case on ground (f), the landlord succeeded even though he had not yet decided his precise scheme of redevelopment, but the property was ripe for redevelopment and his evidence was that he would implement the most profitable scheme (*Yoga for Health Foundation* v *Guest* [2002] EWHC 2658, ChD). However, most of the other reported judgments on ground (f) have required a greater fixity of intention in relation to the details of the scheme. Usually a landlord will have a better chance of succeeding on ground (f) if he can demonstrate a detailed development plan for which planning permission has been granted or is very likely to be granted and also demonstrate that any necessary finance would be available. Although not strictly essential, evidence of the likelihood of the grant of planning permission might be given, for example, by a planning consultant.

Where the landlord is a company, it is also usual to put in evidence the minutes of a resolution of the board determining that the proposed scheme is to be implemented subject to obtaining possession of the premises and, where necessary, authorising counsel for the company to offer the court an undertaking by the company that it will carry out the intended development within a specified period from obtaining possession of the premises. That time period should contain a sufficient allowance for obtaining statutory consents, the preparation of working drawings, contract tendering and other preliminaries.

As for the "reasonable prospect" of being able to implement the scheme, it is not necessary for every possible precondition for starting the development (planning permission, passing of building plans, offer of finance, etc) to be in place by the time of the court hearing (*Dolgellau Golf Club* v *Hett* [1998] 2 EGLR 75, CA). The landlord simply needs to show a firm decision to proceed coupled with a likelihood or reasonable prospect of fulfilling any such preconditions. "Reasonable prospect" does not mean "more likely than not" ; it simply means a real chance rather than a fanciful hope (*Cadogan* v *McCarthy & Stone Developments Ltd* [1996] EGCS 94, CA; *Gatwick Parking Services Ltd* v *Sargent* [2000] 2 EGLR 45, CA).

If the redevelopment cannot be carried out until something occurs which is entirely outside the landlord's control, and which is wholly speculative and arbitrary (such as a need to obtain the right to carry out works on land outside the landlord's control), then the necessary prospect of ability to implement the intention may not yet exist. This does not apply in most cases to the need to obtain planning permission for the proposed development, since that is not an arbitrary matter; a planning application must be determined properly having regard to the local development plan, ministerial guidelines and other ascertainable criteria. For this reason, the obtaining of planning permission is not treated as necessarily speculative, but the existence of a recent refusal of planning permission, especially on appeal, may indicate that there is no reasonable prospect of obtaining permission for that redevelopment (*Coppen* v *Bruce-Smith* [1998] EGCS 55, CA).

To fall within ground (f), the landlord must show that "he could not reasonably do [the work] without obtaining possession of the holding". Here "possession" means legal possession as well as physical possession (*Heath* v *Drown* [1973] AC 498, HL). Accordingly, the intended work must go beyond that which the landlord has the legal right to carry out under any powers reserved to him in the existing tenancy (which powers would normally be repeated in any new lease under section 35). So, for example, if there is a power in the present lease for the landlord to enter and alter the premises internally, that would defeat his objecting to a new tenancy on grounds of redevelopment if his proposed works to the tenant's holding were confined to internal alterations.

The landlord must intend, and have a reasonable prospect of being able, to start the redevelopment within a reasonable time from the termination date of the current tenancy (*Edwards* v *Thompson* [1990] 2 EGLR 71, CA; *London Hilton Jewellers Ltd* v *Hilton International Hotels Ltd* [1990] 1 EGLR 112, CA). The termination date will usually be three months and two weeks after the date of the Court Order (see 10.23). However, by section 31(2)(a), the landlord will succeed even if possession for starting the work is not required until a later date, so long as that is still within a year of the termination date specified in the section 25 notice or section 26 request. In many cases, though, the delays inherent in the procedures under the Act will result in the hearing not taking place until after the specified termination date has long passed, so the time extension under section 31(2)(a) may rarely give real benefit to the landlord. If the landlord cannot start the redevelopment within these time periods, the tenant is entitled to a new tenancy and the landlord would have to press for only a very short term to be granted or for the insertion of a redevelopment break clause (see 12.3).

Section 31A of the Act provides two potential mechanisms for a tenant to obtain a new tenancy notwithstanding the landlord's intention to carry out certain works of redevelopment. The first is contained in section 31A(1)(a), and applies where the intended works to the tenant's holding, although sufficiently substantial to fall within ground (f), would not take a substantial time to carry out and would not interfere to a substantial extent with the tenant's use of holding. In such a case, the tenant can offer to include, in any new tenancy, terms for access such that the landlord could reasonably carry out the intended works without obtaining possession of holding. The landlord can oppose this if the carrying out of the works would interfere to a substantial extent with the use of the holding by the tenant for the purpose of his business and/or would interfere for a substantial time with the use of the holding by the tenant for the purposes of his business. These tests are objective; the fact that the tenant may be prepared to put up with substantial interference from the works is of no relevance and does not enable him to waive either or both of these tests (*Redfern* v *Reeves* [1978] 2 EGLR 52, CA). Once again, and for the reasons given above, the works to be considered are those which go beyond anything the landlord may be entitled to do under powers reserved to him under the current tenancy; works he is entitled to do under any such powers (for example, repairs where he has a right of entry to carry out repairs) are to be ignored. The financial effect on the "interference" on the tenant's business is irrelevant. It is the physical interference that is to be assessed, and this includes both interference while the work is carried out and the continuing interference that would arise if the reconstructed premises were different from the holding such that the tenant could not in practical terms resume business in just the area formerly constituting the holding, for example where the result of the work would be that the holding ceases to exist as a separate unit and/or that major work would be needed to re-establish the holding (*Pumperninks of Piccadilly Ltd* v *Land Securities plc* [2002] 2 EGLR 147, CA; *Blackburn* v *Hussain* [1988] 1 EGLR 77, CA).

The second mechanism offered to tenants by section 31A is in section 31A(1)(b), which applies where the intended works mainly affect only part of the holding. The tenant can seek a new tenancy of the remainder if it constitutes an "economically separable" part. For this purpose, "economically separable" means that the aggregate of the rents which, after the completion of the intended work, would be reasonably obtainable on separate lettings of that part and the remainder of the premises affected by or resulting from the work would not be substantially less than the rent which would then be reasonably obtainable on a letting of those premises as a whole. Even if the part that the tenant wishes to remain is "economically separable", the tenant must also show that either:

(a) he is willing to grant the landlord access or facilities in connection with the intended works as terms of the new tenancy, and in the light of these means of access and/or facilities the landlord is reasonably able to carry out his intended works on the remainder of the holding or
(b) without granting any means of access or facilities the landlord is none the less able reasonably to carry out his intended works on the remainder of the holding.

Where the works will fall short of actual demolition of all or most of the holding, the landlord should endeavour to ensure that the nature and extent of the scheme falls within ground (f) and should take legal advice on this when planning the works. He will want to avoid the tenant renewing as to part under section 31A(1)(a) or (b) and should therefore design his scheme and his programme of works so that all parts of the tenant's holding will be affected by structural works to a considerable extent and for a considerable time, and if possible with the result that the holding is no longer identifiable and capable of being relet by itself. Ideally the landlord should put himself in a position to establish intention and ability by producing planning permission, passing of building plans, a title free from restrictive covenants or adverse easements which would preclude the work, evidence of finance, and board resolutions resolving to proceed with the development and authorising counsel to give the court and the tenant an undertaking on behalf of the landlord that work will commence as soon as practicable after possession is obtained.

A landlord who intends to oppose renewal on ground (f) should take advice as to his likely liability to pay statutory compensation to the tenant (see Chapter 14).

Casenotes

Joel v *Swaddle* [1957] 3 All ER 325: The landlord's scheme included removing load-bearing walls and replacing them by spaced-out columns and changing the level of the floor. Held: The landlord succeeded on ground (f).

Bewley (Tobacconists) Ltd v *British Bata Shoe Company Ltd* [1958] 3 All ER 652, CA: The landlord's scheme incorporated altering the lavatory accommodation involving a small amount of demolition and reconstruction, the reconstruction of an entirely new shop front, and demolition of three-quarters of a back wall. Held: The landlord succeeded on ground (f).

Housleys v *Bloomer-Holt* [1966] 2 All ER 966: The tenancy comprised a yard with a wooden garage on one-third of it, and a brick boundary wall. The landlord's scheme involved demolishing the wooden garage and the wall, and laying concrete over the whole yard to create a lorry turning area to serve the landlord's adjacent premises. Held: This did involve demolition of the whole of the "premises comprised in the holding" (ie the garage), and concreting was probably work of construction, so the landlord succeeded on ground (f).

Barth v *Pritchard* [1990] 1 EGLR 109, CA: The tenancy comprised offices on two floors of a building. The landlord's scheme involved resiting a staircase, rewiring, provision of new toilets, installation of central heating, restoration of the main roof, decoration, carpeting and closing of openings in the party wall. Held: The court can only consider works intended to be carried out within the tenant's holding involving the structure. In this case those works were insufficiently substantial to be within ground (f); most of the substantial works were outside the holding.

Gilmore Caterers Ltd v *Governors of St Bartholomew's Hospital* [1956] 1 QB 387: The landlord had entered into an agreement with a third party to grant him a building lease of the premises and adjacent premises for a term of 48 years, the lease to be granted once the third party had cleared the site and built a new building to plans approved by the landlord. Held: The landlord had proved his intention within ground (f).

Spook Erection Ltd v *British Railways Board* [1988] 1 EGLR 76: The landlord entered into an agreement for lease with supermarket operators, under which the latter agreed to redevelop the site and were then to be granted a 99 year lease at a premium. Held: The landlord had proved his intention within ground (f).

PF Ahern & Sons Ltd v *Hunt* [1989] 1 EGLR 74: The landlords (who were the widow and daughters of the original landlord and had little business experience) instructed a surveyor to pursue the prospect of redeveloping the site. The surveyor gave evidence that the site was ripe for redevelopment and could be disposed of by granting a 125 year building lease. The landlords gave evidence that they wanted to maximise the capital return on the site and were content to act on the surveyor's advice. The tenant argued that the landlords had delegated these matters to the surveyor and did not themselves have the necessary intention. Held: The landlord's intention on ground (f) was sufficiently proved.

Betty's Cafes Ltd v *Phillips Furnishing Stores Ltd* [1959] AC 20, HL: Shortly after the longstanding tenant had been granted its latest lease of the premises, a long reversionary lease was granted. Near the end of the lease, the tenant served a section 26 request on the new immediate landlord, who gave a counter-notice opposing a new tenancy on ground (f). The immediate landlord intended to occupy the premises itself after redeveloping, but could not satisfy the five-year rule for ground (g) (see below) and therefore was opposing on ground (f). During the court hearing, the immediate landlord passed board resolutions that if they successfully obtained possession they would spend £20,000 reconstructing the property and that they would give the court an undertaking to do so. Held: The relevant time to assess the necessary firm and fixed intention to redevelop is the date of the court hearing. It did not matter that redevelopment was to be a precursor to the landlord's own occupation. Ground (f) was proved.

DAF Motoring Centre (Gosport) Ltd v *Hupfield & Wheeler Ltd* [1982] 2 EGLR 59, CA: The landlord's scheme was to redevelop the premises in conjunction with adjoining premises already occupied by the landlord. The landlord had not yet obtained planning permission or finance for his scheme but the county court judge was satisfied, on the evidence, that it was likely that planning permission would be forthcoming and that there was a reasonable prospect of obtaining finance from the landlord's bank or from certain other specified sources. Held: It was not necessary for the landlord to come forward with a cut-and-dried scheme. A sufficient intention for ground (f) had been proved.

Capocci v *Goble* [1987] EGLR 102: The landlord obtained outline planning permission for constructing a block of flats on the site. He had talks with a local developer who had the necessary funding agreed in principle by his bank and was willing to start clearance of the site as soon as detailed planning permission and vacant possession were obtained. The rest of the terms of the deal between the landlord and the developer had not yet been worked out and detailed planning permission had not yet been obtained. Held: A reasonable prospect of proceeding with the development had been proved. A new tenancy was refused.

Edwards v *Thompson* [1990] 2 EGLR 71, CA: The planning permission for converting the demised premises and a nearby barn into dwellings, and to construct new houses on adjacent land, was subject to conditions precluding occupation until the entire development and its access had been completed. Although the landlord had arranged finance and a building contract in respect of the demised premises and the barn, the land on which the houses were to be built was not owned by the landlord but by someone who had not yet found a builder or developer or even costed the redevelopment. Held: A firm and settled intention to redevelop had not been shown since it was unlikely that the construction of the five houses would be ready to proceed within a reasonable time of the termination of this tenancy. A new tenancy was to be granted.

London Hilton Jewellers Ltd v *Hilton International Hotels Ltd* [1990] 1 EGLR 112, CA: The hotel company opposed a new tenancy of a shop within their hotel, intending to reconstruct the premises and to include it in the hotel bar. The company passed a board resolution to this effect and gave an undertaking to do the work when possession was obtained. Held: This was decisive evidence of a compelled fixity of intention. A new tenancy was refused.

Heath v *Drown* [1973] AC 498, HL: The existing lease reserved the right for the landlord to enter the premises and carry out necessary repairs to the building. The landlord opposed a new tenancy on ground (f) but merely intended to secure an unsafe front wall by inserting tie-rods through the building and rebuilding part of the front wall. Held: That work could be done if the tenant was given a new tenancy since it would repeat the reservation of those rights. Therefore legal possession from the tenant was not needed in order to do the work even if a certain amount of physical possession was needed. Ground (f) failed and the grant of a new tenancy was ordered.

Redfern v *Reeves* [1978] 2 EGLR 52, CA: A clause in the current lease permitted the landlord to enter and execute structural repairs to the premises. The landlord opposed renewal on ground (f). His scheme involved structural repairs and also alterations. The tenant proposed under section 31A(1)(a) that the landlord should be given a right of access under the new tenancy to carry out those works. Held: The only works to be considered under section 31A were those for which legal possession was required, namely the alterations, not the repairs. Under section 31A(1)(a), the court must assess the likely interference from those works with the tenant's use of the holding; whether or not they would impact on the tenant's business or goodwill was irrelevant. Since the alterations would interfere substantially with the use of the holding for a substantial time, a new tenancy was refused even though the tenant was willing to put up with the interference.

Price v *Esso Petroleum Co Ltd* [1980] 2 EGLR 58, CA: The landlord of a petrol station opposed renewal on ground (f). His scheme involved 16 weeks' work demolishing the existing canopy, building and pumps and erecting new ones in different locations. The existing tenancy reserved rights for the landlord to enter and carry out improvements, additions and alterations and so the tenant argued that the landlord would not require legal possession to do the work. If that was wrong, the tenant would give the landlord access pursuant to section 31A(1)(a). Held: The landlord's proposed works were within his powers under the terms of the tenancy so, following *Heath* v *Drown*, the tenant was entitled to a new tenancy incorporating the same terms. If that had not been the case, the tenant would have failed under section 31A(1)(a) because 16 weeks disturbance would be "substantial" and the tenant's belief that this would not affect his business was irrelevant.

Cerex Jewels Ltd v *Peachey Property Corporation plc* [1986] 2 EGLR 77, CA: The landlord's scheme for renovating a building involved 12 weeks' work affecting the tenant's part, including laying a new floor slab, which would take two weeks, and other works which mainly fell within a provision in the existing lease allowing the landlord to enter and carry out repairs to the building. The tenant offered to take a new tenancy with additional powers for the landlord under section 31A(1)(a). Held: The work which the landlord could not reasonably carry out without possession of the holding would only last two weeks. That was not "a substantial time", so the tenant's offer was accepted and a new tenancy was ordered.

Blackburn v *Hussain* [1988] 1 EGLR 77, CA: The landlord's scheme to integrate three shop units, one of which was let to the tenant, into one open plan area, would affect the tenant's unit for at least three months and it would need to be closed for most of that time. The tenant offered to take a new tenancy on the terms envisaged by section 31A(1)(a). Held: The interference with the tenant's business would be very substantial, so a new tenancy was refused.

Turner v *Wandsworth London Borough Council* [1994] 1 EGLR 134, CA: The landlord opposed renewal on ground (f), intending to demolish the old buildings on the site, tarmac the surface and then let the land for four years to the adjoining owner for car parking, with a view to selling the land for redevelopment afterwards when the market would have hopefully improved. Held: The motive of a future sale did not detract from the immediate genuine intention to carry out work of demolition and construction. Ground (f) was proved.

8.8 Ground (g) — landlord's intention to occupy

> Subject as hereinafter provided, that on the termination of the current tenancy the landlord intends to occupy the holding for the purposes, or partly for the purposes, of a business to be carried on by him therein, or as his residence.

This provision is supplemented by section 30(1A), (1B), (2) and (2A), mentioned below.

For this ground of opposition, the landlord must be able to establish a genuine and firm intention to occupy the holding for a business he intends to conduct there and/or as his residence; this is a question of fact. As in the case of intended redevelopment under ground (f), discussed at 8.7, there must also be a reasonable prospect of having the ability to fulfil that intention.

Again as in the case of ground (f), it is not necessary for every possible precondition for starting the use (eg planning permission) to be in place by the time of the court hearing, nor need it be shown that the intended business was likely to succeed after it has been started (*Dolgellau Golf Club* v *Hett* [1998] 2 EGLR 75, CA). The landlord simply needs to show a firm decision to start carrying on the business in the premises, coupled with either a likelihood or a reasonable prospect of being able to do so — this prospect is to be the subject of a "reality check" to ensure that it is realistic (*Zarvos* v *Pradhan* (2003) 2 EGLR 37, CA).

The immediate intention to occupy apparently need not be in respect of the whole of the tenant's holding, provided that there is an immediate intention to occupy a substantial part with the remainder following (*Method Developments Ltd* v *Jones*) [1971] 1 All ER 1028). However, if the intention is only to occupy an insubstantial part of the holding then it is probable that the landlord will not succeed on ground (g), but an alternative approach might be to serve notice on ground (e) and offer as alternative accommodation the part of the holding not required by the landlord, assuming that it is large enough and suitably located for the tenant to carry on his business there.

As in the case of grounds (d) to (f), this ground requires the landlord to be in a position to give effect to his intention "on the termination of the current tenancy". However, unlike those grounds, the possibility of an extension of time of up to a year under section 31(2)(a) does not apply.

A landlord can use ground (g), rather than ground (f), even if he intends first to make substantial alterations to the property in order to make it suitable for the purpose of his intended occupation (*JW Thornton* v *Blacks plc* [1986] 2 EGLR 61), but not if he intends totally to demolish the existing property (*Nursey* v *P Currie (Dartford) Ltd* (1959) 173 EG 493, CA; *Jones* v *Jenkins* [1985] 1 EGLR 113, CA).

With the exceptions mentioned below, the intended occupation of the holding must be for the landlord himself, for use as his residence and/or for a business to be carried on by him — not for occupation and use by someone else (*Tunstall* v *Steigman* [1962] 1 QB 593). An intention to carry on a lettings business in the premises will not normally fall within ground (g) because the holding will then be occupied by the proposed new tenants and not the landlord (*Jones* v *Jenkins* [1986] 1 EGLR 113, CA).

The landlord's business can be intended to be conducted at the premises using a management company acting as the agent of the landlord, or by the landlord in partnership with another, while remaining within ground (g) (*Skeet* v *Powell-Sheddon* [1988] 2 EGLR 112, CA; *Teesside Indoor Bowls Ltd* v *Stockton-on-Tees Borough Council* [1990] 2 EGLR 87, CA).

Where the landlord is a company in a group, intended occupation by another company in the same group will count as intended occupation by the landlord for ground (g) (section 42(3)). Where the landlord is one or more individuals, intended occupation by a company which he or they control will also count as intended occupation by the landlord for ground (g) (section 30(1A)). Where the landlord is a company controlled by one or more individuals, intended occupation by the individuals will count as intended occupation by the landlord company but not if the individuals acquired their controlling interest in the company less than five years before the termination of the current tenancy and the property has been continuously let, since that acquisition, on a tenancy, or successive tenancies, protected by the Act (section 30(1B), (2A)).

In the case of landlords who are trustees of the property, they may be entitled to rely for ground (g) upon an intention that a beneficiary under the trust will occupy the property (section 41(2)); however, this only applies if the terms of the trust (either expressly or at the trustee's discretion) allow the beneficiary to do so, and he does so under that power and not under some other arrangement (*Frish* v *Barclays Bank* [1955] 2 QB 541, CA; *Carshalton Beeches Bowling Club Ltd* v *Cameron* [1979] 1 EGLR 80, CA).

However, ground (g) cannot be used if the landlord's interest (or a previous intervening interest which has merged in it) was purchased or created less than five years before the termination of the current tenancy and since that purchase or creation the holding has been continuously comprised in a tenancy or successive tenancies protected by the Act. Here "purchased" means bought for money; where that applies, the five year period starts with the date on which the new landlord contracted to purchase his interest. "Created" normally refers to cases where the landlord's interest is leasehold and means the date of creation (ie execution) of the landlord's lease. However, in cases where the landlord's interest (freehold or leasehold) is held by the landlord on trust, except in the case of a trust arising under a will or on intestacy, "creation" of an interest includes the creation of the trust (section 41(2)). This means that where ground (g) relates to a trustee landlord having the intention that a beneficiary of a trust will occupy the premises, the trust in favour of that particular beneficiary must have been created at least five years before the termination of the current tenancy.

For these purposes, the expression "termination of the current tenancy", to which the five year period runs, has generally been taken to mean the termination date specified in the section 25 or section 26 notice, although in *Diploma Laundry Ltd* v *Surrey Timber Co Ltd* (1955) 166 EG 68, CA it was taken to mean the date when the landlord issued his opposition on ground (g) in a section 25 notice or in a section 26(6) counter-notice.

The application of the five year rule to any factual situation requires careful attention to the wording of section 30(2) and has given rise to numerous court cases. It does not, for example, prevent a landlord from opposing a renewal under ground (g) if he buys a property with vacant possession and then grants a tenancy for say three years to a new tenant and remains the landlord, even if he will only have owned the property for under five years by the termination date. However, if he sells the property during the term of that tenancy, the new landlord will not be able to oppose a renewal under ground (g) because the property will have been let on a tenancy protected by the Act throughout his period of ownership and that period will have been less than five years.

The courts have tended to apply common sense so far as possible to section 30(2) so, for example, a succession of leasehold interests held by the same landlord may be treated as a continuous interest, and transfers of the reversion to and from trustees may be ignored and continuous beneficial ownership assumed.

As for cases where a change of competent landlord occurs between the giving of the section 25 notice (or counter-notice for a section 26 request) and the court hearing, in circumstances where this does not infringe the five year rule, see 8.9.

Casenotes

Expresso Coffee Machine Co Ltd v *Guardian Assurance Co* [1958] 2 All ER 692: The landlords passed board resolutions expressing an intention to occupy the premises for the purpose of their business if they successfully obtained possession. However, they were also interested in the possibility of acquiring a nearby property which would give better accommodation, if terms could be agreed with the owners. Held: The present intention was real and fixed, even if it transpired at a future time that the occupation was short-lived due to a later move. Ground (g) was proved.

Gregson v *Cyril Lord* [1962] 3 All ER 907, CA: The landlord opposed a new tenancy on ground (g) intending to use the premises as offices. The county court judge refused to consider the question whether this required planning permission, and held that ground (g) was proved. Held: The judge must consider whether there is a reasonable prospect of the landlord being able to implement his intention. (The case was resubmitted for a new trial.)

Westminster City Council v *British Waterways Board* [1984] 2 EGLR 109, HL: The council was the tenant of land it used as a refuse truck depot. The board, as landlord, wanted to use the site in conjunction with its adjacent canal to provide a recreational facility, and opposed a new tenancy on ground (g). This proposal required planning permission for change of use and the council (who were the local planning authority) argued that they would not grant it. Held: There was a reasonable prospect of the landlord obtaining planning permission on appeal to the Secretary of State. Ground (g) was proved.

Cox v *Binfield* [1989] 1 EGLR 97: The landlord's intentions for living in part of the property and operating a business from another part were ill thought out and likely to fail after a period of time. Held: Nevertheless, the landlord intended to carry them out, and they were not totally unrealistic. Ground (g) was proved.

Europarks (Midlands) Ltd v *Town Centre Securities plc* [1985] 1 EGLR 88, Ch D: The tenancy comprised a multi-storey car park which served the landlord's adjacent large development. The landlord opposed a new tenancy on ground (g), intending to run the car park itself, and produced board resolutions, quotations from suppliers of car park control equipment, and an affidavit from a director as to the settled intention. Held: Ground (g) proved.

Skeet v *Powell-Sheddon* [1988] 2 EGLR 112, CA: The tenancy comprised a private hotel. The landlord opposed a new tenancy on ground (g) intending to enter into partnership with her businessman husband and run the hotel with her daughter, who was studying hotel management. Held: Ground (g) was satisfied and a new tenancy was refused.

Teesside Indoor Bowls Ltd v *Stockton-on-Tees Borough Council* [1990] 2 EGLR 87, CA: The council as landlord of a bowling club opposed a new tenancy on ground (g), intending to operate the same type of business using a suitably experienced management company. The council would retain the right to make policy decisions and would be advertised as the provider of the facilities. Held: Ground (g) was satisfied. A new tenancy was refused.

Meyer v *Riddick* [1989] NPC 66 CA: Three people were the joint landlords. They held the freehold as tenants in common, each owning one-third. They opposed their tenant's application for a new tenancy on ground (g). The intention was that occupation would be taken by a firm of solicitors in which two of the three joint landlords were partners; and that, in order for the firm to have exclusive possession and be entitled to exclude the third joint owner (who was not a solicitor), the firm would be granted a commercial lease of the premises by the three joint landlords. Held: This was not within ground (g) and the tenant was entitled to a new tenancy.

HL Bolton (Engineering) Ltd v *TJ Graham & Sons Ltd* [1957] 1 QB 159, CA: The landlord let premises to the tenant who sublet part, terminable on three months' notice. The tenant surrendered his tenancy to the landlord without payment and the landlord then gave a section 25 notice to the subtenant, opposing a new tenancy on ground (g). Held: The relevant interest was the immediate reversion, ie the head tenancy, and while it had been surrendered and merged in the freehold it had not been "purchased" by the landlord so the five year rule was not infringed.

Frederick Lawrence Ltd v *Freeman Hardy & Willis Ltd* [1959] Ch 731, CA: In 1933 the tenant sublet a property for a term ending on 25 March 1959. On 10 March 1954 he was granted a new 99 year lease of the property. On 30 June 1954 he agreed to sell his goodwill and premises to an assignee, but no part of the price was specified as the consideration for the transfer of the lease. The transfer took place on 1 November 1954. The assignee served a section 25 notice on the subtenant specifying a termination date of 25 March 1959. Held: The leasehold interest was "created" on 10 March 1954, when the lease was executed, and this was more than five years before the expiry date of the notice. However the assignee had "purchased" the lease, since its acquisition was part of the single larger transaction, and he had done so on 30 June 1954, which was less than five years before the termination date. The five year rule was infringed. A new tenancy was ordered.

Artemiou v *Procopiou* [1966] 1 QB 878, CA: The landlord acquired his first leasehold interest in the property in 1960, and was granted a further lease in 1963. He sublet the premises for two years from 1963. In 1965 he opposed the subtenant's application for a new tenancy on ground (g). Held: The landlord acquired his interest in 1960, not in 1963, so the five year rule was satisfied.

Northcote Laundry Ltd v *Frederick Donnelly Ltd* [1968] 1 WLR 652, CA: On 23 September 1964 the tenant was granted a lease of premises for a term commencing on 29 September 1964. The landlord retained possession of part and was granted a yearly subtenancy from 29 September 1964. The tenant served a section 25 notice on the subtenant in November 1966. Held: The tenant satisfied ground (g). His interest in the premises was created on the date of execution of his lease, not when he went into possession. Since that date (23 September 1964), the sublet part had not continuously been occupied under a tenancy protected by Act since the "subtenant" had been in possession as freeholder for the first six days.

Morar v *Chauhan* [1985] 2 EGLR 137, CA: Between 1968 and 1972 the freehold was vested in the landlord. He transferred it in 1972 to trustees on trust for himself. In 1979 they transferred it back to him and he immediately executed a declaration of trust in favour of his children. In May 1983 he gave a section 25 notice expiring in November 1983 opposing a new tenancy on ground (g), intending that he would occupy the premises for a cleaning business. The tenant contended that the landlord's interest was "created" in 1979, less than five years before the expiry of the notice. Held: The intention was for the landlord himself to occupy, rather than the beneficiaries, so section 41(2) did not apply.

8.9 Change of landlord before hearing date

In those cases where the landlord is opposing a renewal on one of the grounds which involve the landlord intending to do something (grounds (f) and (g)), the question arises how this intention is tested in those situations where a change in the person who is the competent landlord takes place between the giving of the section 25 notice (or counter-notice for a section 26 request) and the court hearing. This may arise, for example, where the landlord who gives the notice or counter-notice then transfers his interest in the premises to a new landlord before the hearing, or in the case of a subtenancy where a superior landlord takes over as competent landlord after the notice is given (see 5.1).

The question is whether the "landlord" whose intention is being tested under ground (f) or (g) is the person who gave the notice (or counter-notice) to the tenant or the person who is the tenant's competent landlord at the date of the hearing, or is both of them.

According to cases decided on grounds (f) and (g) before the 2004 amendments, if a change of competent landlord occurred between the giving of the section 25 notice (or counter-notice for a section 26 request) and the court hearing, the "landlord" whose intention was tested was the person who was the landlord at the date of the court hearing. It was irrelevant whether he was or was not the person who gave the notice or counter-notice or whether the person who gave it had had or had not had the necessary intention at the date of the notice or at any other time (*Morris Marks Ltd* v *British Waterways Board* [1963] 1 WLR 1008 — on ground (f); *Wimbush & Son Ltd* v *Franmills Properties Ltd* [1961] Ch 419, Ch D — on ground (g)). One reason given for this rule was that the notice or counter-notice had (by section 25(6) or section 26(6) as then worded) to state on which of the grounds the landlord "would oppose" an application by the tenant for a new tenancy — this meant would oppose a court application if and when the tenant made one and the court heard it. The matter therefore had to be looked at only at that time.

Following the 2004 amendments, the landlord's statement in the notice or counter-notice is now instead to be whether he "is opposed to the grant of a new tenancy". This appears to refer to his state of mind at the date when he makes the statement. It seems clear that this change of wording was made simply because of the new possibility that the court application might be made by the landlord (seeking a termination order without renewal) rather than the tenant. Nothing in the Government's public consultation exercise leading to the 2004 amendments indicated any intention to change the substance of the law relating to opposition to renewal. It is also likely that the import of the old wording on this issue was not appreciated by those who drafted the 2004 amendments. It may therefore be that the 2004 amendments will be held not to have changed the law as to the person whose intention on ground (f) or (g) is tested.

Where ground (g) is being used to oppose renewal, a sale of the landlord's interest after the giving of the section 25 notice or the section 26 counter-notice would normally infringe the five year rule for ground (g) (see 8.8) if the tests under ground (g) were still applied at the date of the hearing rather than at the date when the notice or counter-notice had been given, and this would lead to a denial of ground (g), but there may be cases of a change of competent landlord which do not infringe that rule and where these considerations will arise.

Casenotes

Morris Marks Ltd v *British Waterways Board* [1963] 1 WLR 1008: A subtenant served a section 26 request on his immediate landlords seeking a new tenancy. They gave a counter-notice opposing a new tenancy on the ground that the freeholder intended to demolish and reconstruct. They then surrendered their lease to the freeholder. The tenant argued that the counter-notice was bad as the person who gave it did not have the necessary intention to redevelop, and that the freeholder who was now the respondent in the proceedings and had the necessary intention was not the competent landlord when the counter-notice was give. Held: The counter-notice was valid and the new landlord could rely on it. A new tenancy was refused.

Wimbush & Son Ltd v *Franmills Properties Ltd* [1961] Ch 419, Ch D: The landlord gave the tenant a section 25 notice stating that a new tenancy would be opposed on ground (g) and then granted an overriding lease to M, who thereby became the tenant's immediate landlord. The tenant argued that the new immediate landlord could not rely on ground (g) specified in the original (now superior) landlord's notice. Held: He could rely on it, since the intention is to be proved at the date of the hearing by the person who is then the landlord.

8.10 Landlord's change of mind and misrepresentation

If a landlord has given a section 25 notice, or a counter-notice to a section 26 request, in which he states that he is opposed, on one of the grounds in section 30, to giving the tenant a new tenancy and then the landlord changes his mind (for example, if he abandons or defers an intended redevelopment of the property), the landlord will often want to withdraw his opposition before the tenant vacates or the court hears the case. The landlord may notify the tenant of his change of mind and, in any court proceedings brought by the tenant, may formally conceded that he cannot now make out the ground or grounds of opposition that he was formerly promoting. However, he cannot actually withdraw the notice in which he stated his opposition to renewal and cannot force the tenant to renew the tenancy if the tenant decides to vacate. As to the effect on the tenant's right to compensation for non-renewal in these circumstances, see 14.1.

The tenant who vacates where the landlord has opposed the grant of a new tenancy and has not withdrawn that opposition may be entitled to claim compensation if the landlord secured the termination of the tenant's tenancy by misrepresentation (section 37A). Such a misrepresentation might be where the landlord gave untrue evidence of tenant's defaults within grounds (a) to (d) of section 30(1) or gave untrue evidence of the facts or intentions constituting grounds (e) to (g) (see the previous parts of this chapter). This can arise in two situations.

The first situation is where the court refuses a tenant's application for a new tenancy made under section 24(1), or grants a landlord's application under section 29(2) for an order for termination of the current tenancy without the grant of a new tenancy, and it subsequently appears to the court that the court's determination was obtained by the landlord having made a misrepresentation or having concealed material facts (section 37A(1)). The second situation, which applies only to cases to which the 2004 amendments apply (see Chapter 1), is where the tenant refrains from making an application under section 24(1) for a new tenancy or else he makes such an application but then withdraws it, and in either case he did so by reason of the landlord's misrepresentation or the concealment of material facts (section 37A(2)). Unlike the compensation provisions of section 37(1C) (see 14.1), these provisions do not — possibly due to an oversight by the Parliamentary draftsman — encompass the situation where, by misrepresentation, the landlord induces the tenant to consent under section 29(6) to the withdrawal of the landlord's application to the court made either under section 24(1) or section 29(2).

The remedy available to the tenant under section 37A is an award of compensation for damage or loss sustained by the tenant as a result of the court determination or of the tenant voluntarily quitting the holding.

However, if the landlord presents his opposition to renewal on, say, ground (f) or (g) on a true and genuine basis but then genuinely changes his mind and does not proceed with the occupation or redevelopment that he had intended to carry out, this would not give rise to liability for compensation, since no misrepresentation will have been involved. Such a genuine change of mind could result from some unforeseen factor such as a major change in market conditions or personal circumstances. The court would have to be satisfied that this change of mind was indeed genuine and only took place after the court determination or the tenant quitting.

Dealing with Subtenants

9

9.1 Steps for tenants to take

Where a tenant has sublet the whole of his premises to a subtenant for a term which is substantially the whole remainder of the term of his own lease, perhaps reserving a nominal reversion of just a few days, the tenant himself will normally have no rights of renewal under the Act and will have to let his own landlord deal with the subtenant in relation to the Act where the subtenant has the protection of the Act. The tenant should not forget, however, to consider issues such as dilapidations (see 9.3).

Where a tenant has sublet the whole or part of his premises but has reserved a reversion of more than 14 months, he should deal with the subtenants in his capacity as their landlord in the normal way (see Chapters 5 and 6).

A tenant who has sublet part of his premises without reserving such a long period of reversion but who remains in occupation of part of the premises for business purposes should also, in his capacity as landlord of the subtenants, take the steps mentioned in Chapters 5 and 6 *vis-à-vis* those subtenants. However, he may only have a small window of opportunity to serve a section 25 notice on them because he can only be certain of remaining their competent landlord for the few days by which the sublease is shorter than the head lease — he remains competent landlord as against his subtenants only until he himself has received a section 25 notice or serves a section 26 request (see 5.1). The Act calls a tenant in this position the "mesne landlord" of the subtenant. If he does not take those steps against the subtenants in time, he may find himself renewing his own lease as to the whole premises, and having to start paying an increased rent (interim or final) before the date when he is able to obtain a rent increase from his subtenants.

For example, if the mesne landlord's own lease ends on 24 December 2006 and he has sublet part for a term ending on 20 December 2006, he can serve a section 25 notice on his subtenant from 20 December 2005 onwards (under the twelve months' rule in section 25), but from 24 December 2005 onwards he might be served with a section 25 notice by his own landlord and, as soon as that happens, his own landlord takes over as competent landlord of the sub-tenant. The mesne landlord's window of opportunity for serving a section 25 notice on his sub-tenant may only be 20 to 23 December 2005 in that situation.

If the mesne landlord omits to serve a section 25 notice on his sub-tenant before losing his capacity as the subtenant's competent landlord, he could ask his superior landlord to serve a section 25 notice on the subtenant. The superior landlord is not obliged to comply with such a request, but if he declines to comply then, in cases where the mesne landlord occupies other parts of the premises and is renewing as to the whole premises (under section 32(2) — see 12.1), the mesne landlord might be able to obtain a reduction in his new rent and interim rent to reflect the delay he will suffer in being able to obtain a rent increase from the subtenant (see12.6).

If a mesne landlord, during the last 16 months of his own contractual tenancy, does serve a section 25 notice on a subtenant while he is still the competent landlord, he must inform the superior landlord "forthwith" and give him a copy of the notice, which he in turn must pass on to any superior landlord higher up the chain (schedule 6, para 7). If the mesne landlord delays giving the superior landlord a copy of the section 25 notice and so deprives him of the opportunity to withdraw it (see below), the mesne landlord might be sued by the superior landlord for damages for breach of statutory duty if he can demonstrate that he has suffered loss as a consequence. Similar rules apply where a mesne landlord receives a section 26 request for a new tenancy from his subtenant.

If a mesne landlord who has served a section 25 notice on a subtenant then ceases to be the competent landlord (usually by receiving a section 25 notice from the superior landlord), he should notify the subtenant of the change of competent landlord (see *Shelley* v *United Artists Corporation Ltd* [1990] 1 EGLR 103, CA).

9.2 Steps for landlords to take

A superior landlord becomes the competent landlord of the subtenants, in place of their immediate landlord (the "*mesne* landlord"), at different times depending whether or not the *mesne* landlord has the protection of the Act. Where the immediate tenancy is not protected by the Act (eg where the immediate tenant is no longer in occupation), the superior landlord becomes the competent landlord of the subtenants 14 months before the term expiry date of the immediate tenancy. Where the immediate tenancy is protected by the Act (eg where the tenant has sublet part but remains in occupation of another part), he becomes their competent landlord once he has served a section 25 notice on the mesne landlord or has received a section 26 request from him.

The superior landlord must give a section 25 notice to any protected immediate tenant before he can give notice directly to any of that tenant's subtenants. The termination date to be inserted in a notice to a subtenant has been considered at 6.5.

If the superior landlord gives section 25 notices to his immediate tenant and to a subtenant on the same day, he should do so in that sequence, although in such cases service in the correct sequence will be assumed unless there is evidence to the contrary (*Keith Bayley Rogers & Co* v *Cubes Ltd* (1975) 31 P&CR 412). However, it is safer practice to give the notices sequentially on successive days.

Before serving a section 25 notice directly on a subtenant, a superior landlord who has become their competent landlord can seek the consent of the mesne landlord, which consent cannot be unreasonably withheld (schedule 6, para 4). If the competent landlord serves the notice without obtaining the mesne landlord's consent, the superior landlord must compensate the mesne landlord for any loss he suffers as a consequence.

If, within two months of the mesne landlord giving a section 25 notice to a subtenant, a superior landlord becomes the subtenant's competent landlord (usually by serving a section 25 notice on the mesne landlord), the superior landlord can within those two months, but not later, notify the subtenant that he withdraws the section 25 notice that had been given to the subtenant and he can serve a fresh section 25 notice if he wishes (schedule 6, para 6). Form 6 of the prescribed forms of notice must be used to notify the withdrawal. This procedure is essential where, for example, the tenant has given a section 25 notice to the subtenant stating that he will not oppose a renewal of the sublease but the superior landlord wishes to oppose renewal. If the superior landlord fails to put himself into a position in which he can withdraw the section 25 notice given to the subtenant and serve a fresh one within the time-limit mentioned, the superior landlord will be bound to grant to the subtenant a new tenancy of the subtenant's holding even if he otherwise could have opposed a renewal on one or more of the grounds of opposition under section 30.

As mentioned above, if a section 26 request for a new tenancy is served on the mesne landlord by the subtenant, the tenant must "forthwith" pass a copy of the request to the superior landlord (schedule 6, para 7). If that is done promptly, it will give the superior landlord an opportunity to serve a section 25 notice on the mesne landlord and then to serve on the subtenant, within two months from the service of the section 26 request on the mesne landlord, a counter-notice under section 26(6) opposing the grant of a new tenancy. If the mesne landlord delays giving the superior landlord a copy of the section 26 request and so deprives him of the opportunity to serve the counter-notice in time, the mesne landlord might be sued by the superior landlord for damages for breach of statutory duty if he can demonstrate that he has suffered loss as a consequence.

A landlord should remind any tenant who has sublet about the obligations under para 7 of schedule 6, he should monitor the situation, and he should serve his section 25 notices on the tenant and subtenant at the earliest possible times.

9.3 Dilapidations in cases of subletting

Where no new lease is granted and the landlord retakes possession of the premises, he will ordinarily be able to claim damages from the outgoing tenant if the premises are handed over in a poorer state of repair than that required of the tenant under the terms of the lease which has terminated. The assessment of such damages is a complex question and is outside the scope of this book, but generally the landlord cannot recover in damages more than the amount by which the value of his reversion is diminished by the disrepair (Landlord and Tenant Act 1927, section 18, and see 4.12).

Where a tenant does not renew his lease of premises which are fully or partly sublet to subtenants who themselves have the protection of the Act, the subtenants may obtain new tenancies from the superior landlord in his capacity as their competent landlord (see 5.1). In *Family Management* v *Gray* [1980] 1 EGLR 46, the landlord in that situation failed to recover any damages for disrepair from an outgoing tenant. The subtenant of the whole property was being granted a renewal of his sublease, which had been on full repairing terms. Those terms were to be repeated in the new lease and the new rent was fixed on the assumption that the full repairing obligations had been fully observed (see 12.6 and 12.10). Consequently the landlord had suffered no reduction in rent and would have a new tenant (namely the former subtenant) under full repairing obligations; therefore the landlord was treated as suffering no diminution in the value of his reversion.

In some cases, however, there may be a difference between the repairing obligations owed by the outgoing tenant to the landlord and the repairing obligations of a subtenant under his sublease. For example, the outgoing tenant's lease may have been on full repairing terms but the sublease might have the subtenant's repairing liability limited by reference to a schedule of condition. When the subtenant obtains a new lease under the Act, he will normally seek to have repairing obligations in his new lease in terms no more onerous than those contained in his old sublease and he will be relying on *O'May* v *City of London Real Property Co Ltd* [1983] 1 EGLR 76, HL to justify this stance (see 12.4). The landlord might have to grant a new tenancy to a person who was formerly a subtenant incorporating repairing obligations which are less onerous than those of the outgoing tenant. This is why many landlords insist that the terms of subtenancies must be similar to the terms of the lease out of which they are granted or else must be contracted out of the Act. The opportunity for the landlord to achieve that protection will not apply where the tenant grants a sublease is granted in breach of the terms of his own lease (even unlawful subtenancies may have the protection of the Act — see 2.4), but it may possibly be the case that, in that situation, the courts will not apply the *O'May* principles quite as strongly as in more usual situations.

A further difficulty may arise where the outgoing tenant vacates part of the premises and a subtenant obtains a new full repairing lease of the other part. The landlord may be able to recover damages for disrepair from the outgoing tenant in relation to the part vacated. However, the landlord may be under a duty of care to ensure that any repair works to that part are carried out causing as little disturbance as possible to the subtenant, and this may increase the cost and duration of the works. This may affect the full recovery of the costs from the outgoing tenant.

Where the sublet part is in disrepair, in breach of the subtenant's repairing obligations under his former sublease, the landlord might consider opposing a renewal by the subtenant on ground (a) of section 30(1) of the Act (see 8.2). If he does not, he should try to obtain a covenant in the subtenant's new lease requiring the necessary repair works to be carried out within a specified time, since the landlord may not be able to recover the cost of repair from the former tenant for the reasons considered in the *Family Management* case (above).

Court Applications and Procedure 10

10.1 Provisions governing court applications

The reasons why a court application may have to be made has been explained at 1.4, 4.1 and 4.8. Unless the landlord and the tenant agree otherwise, as mentioned below, the application to the court must be lodged at court no later than the expiry date specified in the landlord's section 25 notice, or the day before the expiry date specified in the tenant's section 26 request (section 29A). This time-limit is called the "statutory period".

The parties may agree to extend the "statutory period" (section 29B). Their agreement to extend it must be recorded in writing (section 69(2)). Further agreements to extend it are permissible, provided they are made before the expiry of the last agreed extension (section 29B(2)).

In claims where the landlord is not opposing renewal, either party may make the application to the court to determine the terms of renewal (section 24(1)). If, on the other hand, the landlord is opposing renewal, the tenant may apply to the court for an order renewing the tenancy (section 24(1)) or the landlord may apply to the court for an order ending the current tenancy without any renewal (section 29(2). In the latter case, if the landlord is unsuccessful in opposing the renewal, the claim then proceeds as if there were an application for an order renewing the tenancy (section 29(4)(b)).

The party commencing the claim is referred to in the proceedings as "the claimant". The other party is the "defendant".

Any application by the landlord may only be withdrawn if the tenant agrees (section 24(2C); section 29(6). This rule is needed because otherwise the tenant's rights of renewal could be lost. The tenant however may require the court to dismiss a landlord's application for a renewal of the tenancy if the tenant decides not to seek a new tenancy (section 29(5)).

In order to avoid duplication of court applications, sections 24(2A) and 24(2B) provide that, once a party has commenced court proceedings under section 24(1) or section 29(2) and has served those proceedings on the other party, that other party is barred from making his own application under section 24(1). It was plainly intended that there should be a similar rule that the landlord cannot apply to the court under section 29(2) for a termination order once either party has made and served on the other an application under section 24(1). However, there is an unfortunate lacuna in section 29(3) which simply bars the landlord from applying to the court once an application under section 24(1) has been made, but it is silent as to the landlord having to have been served with that application. The words about service which appear at the end of sections 24(2A) and 24(2B) do not appear in section 29(3). Consequently a landlord may inadvertently infringe section 29(3) if he makes an application under section 29(2) not realising that the tenant has already made a section 24(1) application which has not yet been served on him and he is therefore unaware of it. Presumably the court would strike out the later application as being a nullity, if either party asked the court to do so.

10.2 Civil Procedure Rules

All court proceedings under the Act are governed by the Civil Procedure Rules, commonly known as the CPR.

Part 56 of the CPR applies to all claims under the Act . The rules differentiate between unopposed claims (renewals where only the new rent or new lease terms are in issue) and opposed claims (where the landlord opposes the grant of a new lease on one or more of the grounds set out in section 30(1) of the Act). Unopposed claims are also known as standard lease renewal proceedings. Opposed claims are sometimes called contested lease renewal proceedings. Different procedures must be used: where the grant of a new lease is opposed the procedure used is that set out in CPR 7. For unopposed claims the procedure is that set out in CPR 8.

10.3 The appropriate court

Claims should be started in the county court for the area where the demised premises are situated. Only in exceptional circumstances should the claim be brought in the High Court. "Exceptional circumstances" include cases where there are complicated disputes of fact or if there are points of law of general importance. The list is not exhaustive. The fact that the demised premises are highly valuable, or situated in a development or building of high value, will not of itself usually justify starting the proceedings in the High Court. If a claim is incorrectly brought in the High Court, the court has the power to strike the claim out or transfer it to the appropriate county court. Claims properly brought in the High Court are heard in the Chancery Division. If the claimant believes the claim ought to be heard in the High Court a certificate will need to be filed with the claim form explaining the reason(s) why the claim is considered fit for the High Court.

10.4 Basic contents of the claim form

In order to comply with the CPR, the claim form itself must contain the following particulars: the address of the premises; details of the current tenancy (including date, parties, rent and the length of the tenancy) as well as the date and mode of termination; details of each notice/request served under section 25 and section 26 of the Act; and the expiry date of the statutory period under section 29A(2) of the Act, or any agreed extended period under section 29B(1) or 29B(2) of the Act (see 10.1).

The parties must be the "tenant" and the competent "landlord" (section 44) (see 5.1). Generally, the proceedings will be a nullity if brought by, or made against, someone other than the correct person. The tenant's application must seek a new tenancy in respect of the premises which are comprised in the "holding" (section 32) (see Glossary). The application must state whether there is any part of the demised premises which is not occupied in this way. A court application which mis-states the holding will generally be a nullity unless the court grants the tenant leave to amend the application to refer to the correct holding. Prior to the CPR there were many cases where tenants sought, under the old Rules of the Supreme Court, to amend incorrect applications under the Act. The matter is now governed by the CPR and those earlier cases are no longer decisive. The CPR contains provisions allowing the court to permit by order the correction of mistakes of procedure. CPR 3.10 allows a court to make an order to remedy an error in proceedings. CPR 19.2 allows the court to order a change in the parties to proceedings and CPR 19.5 allows this to happen even after the expiry of the relevant limitation period in certain situations. It has been held recently that the court may amend the name of a party in a case of obvious mistake or in other appropriate cases where material prejudice has not been caused (*Parsons* v *George* [2004] 40 EG 150, CA).

Casenote

Parsons v *George* [2004] 3 EGLR 49, CA: The landlord died and her executors served a s.25 notice on the tenant. The executors then transferred the property to the beneficiary under the will, who was one of the two executors. The tenant's solicitor was informed of the transfer but erroneously named the two executors as landlord in the tenant's court application. When this was pointed out, after the time-limit for applying to the court had expired, the tenant applied to the court for consent to amend the application. Held : There was nothing in section 29 of the Act which prohibited changes to a party after expiry of the time-limit imposed by that section for applying to the court. The requirements of CPR 19.5 were satisfied because the tenant had always intended to claim the new tenancy from the competent landlord and the mistake fell within that paragraph. The court exercised its discretion to allow the substitution — the error was obvious and the correction would not cause the competent landlord any true prejudice.

10.5 Tenant's claim forms

Where the tenant is bringing a claim the claim form must also include the tenant's proposals for the new lease (the rent, the duration of the term, and the other terms of the lease), details of the nature of the business conducted by the tenant, whether any part of the premises is not occupied by the tenant or a person employed by the tenant, and whether the tenant relies on section 23(1A) (controlled companies — see 2.8) or section 31A (redevelopment during new lease — see 8.70) or section 41 (trusts — see 2.10) or section 42 (group companies — see 2.7) and , if so, the basis of that reliance.

Where the premises are occupied by a partnership, as discussed in Chapter 2, it is the "business tenants" who have the right to seek a new tenancy, to be granted to themselves without those who have left the partnership (section 41(5)). The court has the power to order that the new tenancy is it be granted to them together with any other partners, and subject to conditions as to guarantors and otherwise, as the court thinks equitable having regard to the fact that the former named partner or partners will not be tenants under the new tenancy (section 41A(6)).

The claim form has to be served on the landlord (or the landlord's solicitors if, but only if, they have previously confirmed that they have instructions to accept service of proceedings). The same service rules apply as in the case of service of notices under the Act.

10.6 Landlord's claim forms

Where the landlord is making a claim the claim form should contain those details specified in 10.4 as well as (where the claim is unopposed) the landlord's proposed terms for the new tenancy, the name and address of anyone known by the landlord to have an interest in the reversion or the freehold estate, whether the tenancy is one to which section 32(2) of the Act applies (tenant not occupying whole premises — see 12.1) and, if it does apply, whether the landlord requires the new tenancy to be a tenancy of the whole of the property comprised in the existing tenancy or just of "the holding" (see Glossary).

Where the landlord is making an application and the grant of a new tenancy is opposed, the landlord must also set out in full each ground of opposition, full details of each ground and the terms of the new tenancy the landlord would be willing to grant in the event that it was unable successfully to oppose the grant of a new tenancy.

10.7 Acknowledgments and defences

Where the claimant is the tenant and its claim for a new lease is unopposed, the landlord must complete an acknowledgment of service and send it to the relevant county court within 14 days of service. The acknowledgment must set out whether the landlord objects to any of the terms for the new tenancy proposed by the tenant, if so which terms, setting out the landlord's counter-proposals. If the landlord is itself a lessee under a lease with less than 15 year's unexpired term at the date of termination of his tenant's lease, the landlord must state the name and address of any reversioner whose reversion is immediate or will arise within 15 years from that date.

The acknowledgment of service must also state the name and the address of any person having an interest in the property who is likely to be affected by the grant of a new tenancy (such as the landlord's mortgagee) and, where the tenancy is one to which section 32(2) of the Act applies (see 12.1), whether the landlord requires that any new tenancy should be a tenancy of the whole of the property comprised in the tenant's current tenancy .

In cases where the renewal is unopposed and the claimant is the landlord, the tenant's acknowledgment of service must state what type of business is carried on at the premises; whether the defendant tenant is relying on section 23(1A) (controlled companies — see 2.8) or section 41 (trusts — see 2.10) or section 42 (group companies — see 2.7) and , if so, the basis of that reliance; whether any part (and if so which part) of the property comprised in the tenancy is occupied by someone other than the tenant or its employees (such as a subtenant); the name and address of anyone known to the tenant to have an interest in the reversion (whether immediate or within 15 years) on the determination of the defendant's current tenancy and who is likely to be affected by the grant of a new tenancy .

The tenant must also set out its proposed terms for the new tenancy. This must include the new rent the tenant proposes paying, the duration of the new lease and any other amendments to the existing lease (for example, the inclusion of a tenant's break clause).

In cases where the grant of a new lease is opposed by the landlord, the tenant must (whether it is the claimant or the defendant) file a defence or acknowledgment of service in which it must additionally state its proposed terms for a new tenancy (ie rent, duration of tenancy, and whether the tenant is seeking amendment of the existing lease).

In cases where the renewal is opposed and the claimant is the tenant, the landlord must file a defence which must contain the ground or grounds of opposition (these should be recited as they are in section 30 of the Act); the fullest details possible of each ground of opposition (discussed at 10.9 below); if, notwithstanding the opposition to the grant of a new tenancy, one were to be granted the terms of the new tenancy proposed by the tenant to which the landlord objects and the landlord's counter-proposals; the name and the address of any other party having an interest in the property, likely to be affected by the grant of a new tenancy; where the landlord is itself a tenant under a lease having less than 15 years unexpired at the determination of the claimant's current tenancy, the name and the address of the reversioner; if the claimant's tenancy is one to which section 32(2) of the Act applies (see 12.1), whether the defendant landlord requires the new tenancy to be a tenancy of the whole property comprised in the current tenancy (rather than just the holding).

In those opposed claims where the landlord is opposing renewal and is the claimant the landlord's claim must set out the details of the property, the details of any section 25 notice or section 26 request and the same information as would be contained in a landlord's defence (see above).

10.8 Detailing the grounds of opposition

Before the introduction of the CPR, the landlord needed only to recite in the defence the wording set out in section 30 of the Act (see Chapter 8) for the relevant ground or grounds of opposition on which he relied. The CPR however require a landlord who is opposing renewal of the current tenancy to give "full details" of the grounds of opposition. It is considered that for each ground of opposition the bare minimum to be pleaded in the defence is as follows.

Ground (a) — failure to repair

The landlord should append to the defence an up to date schedule of dilapidations, an estimate of the cost of repair and evidence of any damage the disrepair has caused to the value of the landlord's reversion. These document should be referred to in the text of the defence If a notice has been served under section 146 of the Law of Property Act, this should also be referred to.

Ground (b) — persistent arrears of rent

The landlord should append to the defence a schedule of arrears and this should include what has been paid and when, what has not been paid, and a total of the arrears This schedule should be referred to in the text of the defence.

Ground (c) — other breaches of covenant

This ground of opposition is so worded that it cannot include breaches of the tenant's repairing obligations or the obligation to pay rent timeously. So, for example, breach of the user clause, or unlawful subletting, will fall within this ground of opposition and care should be taken to include as much detail as possible (for example the date of the breach in question; whether the breach can be described as a flagrant one; where the breach is one of user, the nature of the unlawful user complained of; where the breach is one of unlawful subletting, the name of the unlawful subtenant, whether a section 146 notice has been served, whether there have been previous unlawful sublettings).

Ground (d) — offer of suitable alternative accommodation

Ideally the landlord will already have written to the tenant describing in detail the accommodation it has offered, in which case the landlord need do no more than append the letter to the defence and refer to it in the text of the defence.

Ground (e) — subletting of part

Here the defence should set out the date of creation of the head tenancy and the subtenancy and should refer to (and explain) the provenance of any deeds of variation, assignment or surrender which affect the original legal statuses of the superior landlord , tenant and subtenant.

Ground (f) — demolition and construction

As set out at 8.7, ground (f) comprises a number of potentially separate grounds of opposition, covering demolition or reconstruction or substantial construction and in each case affecting either the premises comprised in the holding or a part of the holding.

The works the landlord intends to do should be sufficiently described. This is probably best done by identifying, in the defence, those plans or drawings which identify the work intended to be carried out, with the drawings or plans being appended to the defence. Alternatively, if planning permission for the works is required and has been obtained it may be possible to recite details of the intended works from the planning permission.

Ground (g) — landlord's intention to occupy

Here the defence need do no more than recite the landlord's intention as required by the provisions of section 30(1)(g), considered at 8.8.

10.9 Applications for interim rent

These can be brought in three ways. Where the landlord is the claimant, he can include it in his claim form (and vice versa, where the tenant is the claimant) Where the tenant is the claimant, the landlord can include it in his acknowledgment or defence (and vice versa where the landlord is the claimant). Alternatively, the landlord or the tenant may make an application of its own volition under Part 23, CPR. Where the landlord or the tenant applies under CPR Part 23, the claim form must state the address of the property, the particulars of the relevant tenancy, and details of section 25 or section 26 notices or requests.

10.10 Time-limit for serving claims and other documents

The claim form must be lodged at court in accordance with the time-limits set out in 1.1 above. The claim form must be served within two months after the date of issue by the court. The date of issue will usually be a few working days after the original lodgement at court. However it is possible to extend the time for service. The general rule is that any application to extend time must be brought before the two months has elapsed. If the two months has elapsed the court may extend time for service, but only if either the court has been unable to serve the court form or the claimant has been unable to serve the claim form despite having taken all reasonable steps to serve it, and, in both cases, the claimant has acted promptly in applying for an extension.

It should be noted that the permission of the court is usually required where the claimant proposes to serve a claim form on a defendant based outside England and Wales.

10.11 Statements of truth

Claim forms, acknowledgments of service, applications for claims to be commenced in the High Court, defences and witness statements must be verified by a statement of truth in the form prescribed by the CPR. The statement of truth must be signed by the landlord or tenant (as the case may be) or its solicitor. Witness statements must be signed by the maker of the statement.

10.12 Court directions

The CPR oblige all parties to litigation to co-operate with the court to ensure that cases are dealt with "justly" (CPR Part 1). In practice this means that parties must handle lease renewals with a view to avoiding unnecessary expense and dealing with renewals in ways which are proportionate to the amount of money involved, to the importance of the case, to the complexity of the issues and to the financial position of each party. Further, the parties must co-operate with the court to ensure that claims are dealt with expeditiously and have allotted to them an appropriate share of the court's time.

The court assists the parties by taking a more active role in case management. Gone are the days when lease renewals could lie dormant for months on end. Once a defence or acknowledgment of service has been filed, the court will send questionnaires to the parties designed to elicit whether the claim is capable of being tried in a day or less and the amount of expert evidence likely to be called. The court will give directions about management of the claim, including the track (see 16.5).

In unopposed claims, it is usual for the landlord and the tenant to ask the court to defer deciding to which track a claim should be allocated while the parties try and negotiate the new rent, the duration of the new term and other terms of the lease. This process is called "seeking a stay of proceedings" and the court is usually content to grant the parties stays of up to three months, at the end of which period the court expects the parties to inform it whether the negotiations have been successful.

Once the stay of proceedings has expired (and provide the parties have not sought and obtained a fresh stay), or if no stay of proceedings has been granted, the court will be keen for directions to be issued. "Directions" are orders of the court regulating the future conduct of the proceedings and the nature of the directions likely to be given by the court will depend on whether the claim is opposed. Typically directions will include the following.

(a) Permission to substitute or join a new party to the proceedings, in cases where the landlord has sold its reversion after the proceedings have commenced.
(b) The trial of a preliminary issue in opposed claims. The issue in question will be whether the landlord can establish, to the court's satisfaction, that its ground of opposition has been made out (see *Dutch Oven Ltd* v *Egham Estates and Investment Co Ltd* [1969] 1 WLR 1483). All other issues in the claim will be stayed until determination of the preliminary issue. Where the landlord is relying upon more than one ground of opposition the court may conclude that the preliminary issues should be tried separately, but this depends on the facts of each case. For example where the landlord is opposing renewal on the grounds of persistent arrears of rent and also that it intends to demolish and reconstruct the holding, the court might conclude that the first ground of opposition can be quickly and effectively tried in a day while the second ground of opposition would require both parties to call expert evidence in a number of disciplines. On that basis the court might decide to hear initially just the first ground. If, at the end of the trial of that first preliminary issue, the court found in favour of the landlord, the expense of the second preliminary hearing could be avoided.

(c) The trial of a preliminary issue in unopposed claims, for example whether a landlord's section 25 notice was valid.

(d) An order that parties disclose documents (by way of list) relevant to the claim. Under the CPR, it is the duty of both the landlord and the tenant to disclose documents upon which they rely, documents which adversely affect their own case or the other party's case (or which support the other party's case).There is no duty to disclose documents which are legally privileged (eg correspondence about the renewal between the landlord and its solicitor, between a properly appointed expert and the landlord's solicitor, without prejudice correspondence). The duty is not simply one of disclosing documents in the physical possession of the landlords or tenant — it extends to documents they may have given to their solicitors, or as security to their bankers. The list must be verified by a disclosure statement.

In unopposed claims there will be little documentation that needs to be disclosed (often the only relevant documents will be the lease and the notices served under the Act) and the court may even dispense with disclosure if the parties are in agreement. In opposed cases the opposite will apply and there may be an extremely large amount of documentation to disclose, especially in cases where the landlord is opposing renewal on the ground that it intends to demolish, construct or reconstruct the holding. In such a case the landlord might find itself having to disclose its planning permission, any section 106 agreement under the Town and Country Planning Act, documents of title to adjacent land owned by the landlord, documents demonstrating the landlord has the finances available to carry out the works, tenders, statutory consents and the like.

(e) An order that the parties make available for inspection by the other party documents set out in their disclosure list. In practice cases where either party wishes to inspect original documents will be rare. Ordinarily if a party requires sight of a document contained in the other party's list of documents it will ask for a copy of the document or documents.

(f) An order that the parties exchange written statements of witnesses of fact. These will be necessary where the landlord opposes the grant of a new tenancy and in unopposed cases where the parties dispute certain terms of the new lease (eg whether there should be an option to determine the new lease, whether there should be guarantors, the duration of the new lease). The ordinary direction is that there should be simultaneous exchange by both parties. But in opposed cases it is often more sensible for the landlord to serve its evidence first, with the tenant being given an opportunity to serve its evidence in reply at a later date.

(g) An order that the parties be permitted to call expert evidence at trial, subject to the written statements of the experts being exchanged. Again, in opposed cases it is often more sensible for the landlord to disclose its expert evidence first (see also 10.16 and 10.17). The court will insist that the number of experts to be called by each party be limited. This is not usually a problem ,but it can cause difficulties in cases where the landlord is opposing renewal on the ground it intends to demolish, construct or reconstruct the holding, where it is conceivable that the landlord might wish, or be obliged in order to prove its intention, to call experts in architecture, building surveying and town and country planning. The court, under its case management powers, can order the parties to appoint a joint expert (CPR 35.4).This rarely happens in opposed claims. However the court has indicated the general desirability of having a single joint expert ,albeit in a claim involving medical negligence (*Peet* v *Mid-Kent Healthcare Trust* [2002] 1 WLR 211).

In unopposed claims where the only issue in dispute is rent, or the duration of the new lease, or one or more terms of the new lease the most appropriate course is for the landlord and tenant to agree to stay the court proceedings and remit the rent dispute to be dealt with under the Professional Arbitration on Court Terms scheme (commonly known as "PACT" (see 10.15).

(h) An order, in cases where there is a likely to be a large amount of expert evidence, that the parties' experts meet on a without prejudice basis to try and agree facts and identify which issues remain in dispute. Ordinarily the order will require the experts' to record what is agreed in a joint, signed statement

(i) An order that the proceedings be stayed to enable the parties to refer the dispute as to new rent to determination under the PACT scheme (see above).

(j) An order that the landlord's or tenant's application for interim rent (if any) be heard at the same time as the main trial.

(k) An order specifying the date "window" within which the trial will be heard and the likely duration of the trial.

(l) An order permitting a party to amend its claim or defence. While a landlord is not able, for example, to claim a new ground of opposition it might wish to withdraw its opposition to the grant of a new tenancy entirely, or it might wish to drop one of its grounds of opposition. Similarly the parties may wish to amend their case as to the length of the new term they are prepared to accept.

(m) An order that the landlord serve on the tenant (in unopposed claims) a draft lease, with a sequential order that the tenant serve a schedule of its proposed amendments upon the landlord.

Each order will specify the date by which it is to be complied. By agreement, the parties may vary the dates. Sanctions for non-compliance with directions can be severe: a party may be debarred from calling expert evidence, for example. In extreme case its claim or defence may be struck out.

10.13 Withdrawal or discontinuance of proceedings

A claimant may generally discontinue any proceedings without permission of the court. However, there are provisions in the Act preventing a landlord from withdrawing an application he has made under section 24(1) or section 29(2) without first obtaining the tenant's consent to the withdrawal, since doing so might lose the tenant his right to renew (see 10.1).

Where the withdrawal or discontinuance is made more than three months before the termination date under the relevant section 25 notice or section 26 request, that termination date will operate. Where the withdrawal or discontinuance is made later, the tenant's tenancy will continue until at least three months after the date of discontinuance (section 64; and see 10.23).

Discontinuance of the tenant's claim does not affect any right of either party to have interim rent determined; the time-limit for either party to seek an order for interim rent is six months after the end of the continuation tenancy (section 24A(3); see Chapter 11).

Discontinuance automatically obliges the party which has discontinued to pay the other party's legal costs up to the date of discontinuance.

10.14 Mediation

One ethos of the CPR is that parties should only litigate as a last resort. Parties who ignore this are liable to suffer costs penalties. The court expects parties to have exhausted all avenues of negotiation before proceedings are issued. As part of its case management powers, the court may require the parties to existing litigation to remit their dispute to mediation. This is a process whereby a neutral third party, called a mediator, seeks to persuade the parties to resolve their disputes. The court usually makes this order at the stage when it gives directions, but the parties are free to agree to mediate at any time. As mentioned above, a party who declines to refer a dispute to mediation without good cause can find themselves facing costs sanctions, even if they are ultimately successful (see *Dunnett* v *Railtrack plc* [2002] 2 All ER 850, *Hurst* v *Leeming* [2002] EWHC 1051).

Mediators are trained professionals from a variety of disciplines. The parties may agree to appoint a specified mediator, or they may approach one of the many mediation organisations (such as CEDR, which has its own model form of mediation) and ask them to nominate a mediator.

The nominated mediator usually requires both parties to prepare a short written case summary, which should identify the issue or issues in dispute and which should have annexed to it any relevant documents. The case summaries are exchanged. The mediator then convenes a meeting of the parties and their representatives at an agreed venue. It is essential that the parties are represented by a person having authority to reach an agreement which will legally bind the party they represent. The parties will be asked to sign a mediation agreement — this is the document which records their agreement to mediate, names the individuals present, empowers the mediator and sets out the mediators' fee. The parties will be jointly and severally liable for that fee.

The conduct of the mediation is at the discretion of the mediator. Usually the mediator will spend time alone with each party, trying to elicit their objectives in the litigation and possibly explaining the strengths and weaknesses of their case. At some point the mediator will probably convene a meeting between the parties at which he will try and explore common ground between the parties, in order to encourage them to settle the dispute. All discussions, and the case summaries, are confidential and may not be repeated by the mediator or either party.

The advantages of mediation are that it can often lead to a negotiated settlement with which both parties are content (which is useful in matters where there is going to be an on-going landlord and tenant relationship); disputes are resolved quickly, usually on the day of the mediation; if the mediation takes place relatively early in the dispute, it can be significantly cheaper than litigating the dispute at court; everything is confidential; mediation is wholly voluntary and either party may withdraw from the mediation at any time; mediation is also "without prejudice", so that the parties are free to express themselves without fear of any concessions made during the mediation being used against them at any future revival of the litigation; a mediation settlement is only legally binding when reduced to writing and signed by both parties; mediation takes place at a time and venue of the parties' choice; mediation can be used by a party to test the strength of its own case, and the strength of the opponent's case, before a neutral third party; and the majority of cases which are mediated are settled at the mediation stage.

The disadvantages of mediation are that either party can withdraw from the mediation at a time when the other party wishes to continue with mediation; mediation can be used by one party not only to test its case, effectively using the mediator as a sounding board, but trying to test the resolve of the other party and in the long run this might prevent settlement rather than achieve it; and much depends on the qualities of the mediator — is he robust enough, or too robust, and does he fully understand both sides of the dispute?

The property industry tends to be conservative and many people in the industry will find it difficult to see how mediators can assist in resolving disputes under the Act. The orthodox view will be that in unopposed claims the parties will each have received advice from their surveyors as to rental values, and from their legal advisors as to settlement possibilities and tactics.

In favour of the orthodox view it is, of course, doubtful what weight can be attached to a mediated settlement. That will be an important consideration to a landlord hoping to use a rent crystallised on one renewal as evidence in a forthcoming renewal. In opposed cases the orthodox view will be that the position of the parties will probably be too entrenched to be resolved by a third party whose purpose is to resolve the dispute, rather than provide an answer to the dispute. But there are types of claim which can usefully be mediated, such as those unopposed claims where issues other than rent are in dispute. In such circumstances it would be feasible and cost effective for the parties to agree to have the issues mediated before a legally trained and suitably experienced mediator. Further, the mediator enjoys a flexibility not enjoyed by the court; for example the mediator might persuade the parties to enter into a contracted out lease, the court has no power to order this without the consent of the landlord and the tenant.

10.15 Professional Arbitration on Court Terms ("PACT")

The PACT scheme is jointly run by the Law Society and the Royal Institution of Chartered Surveyors (RICS). The scheme currently only applies to unopposed claims under the Act. Under the scheme a solicitor or surveyor will be appointed to act either as an arbitrator or independent expert, which the parties decide at the outset. The differences between an arbitrator and an expert are discussed in below.

Usually the issues which will be referred to the expert or arbitrator will be the amount of the new rent and interim rent, the duration of the new lease, the other terms of the new lease (for example whether there should be a landlord's option to determine) and the drafting of the new lease. The attraction of the scheme is that county court judges, who hear a variety of case involving different areas of the law, are not best suited to deciding the rent under a new lease. Similarly the parties are probably better advised referring difficult drafting issues to solicitors who specialise in the subject and have day to day experience of drafting leases and negotiating lease renewals. Further in cases where a site visit to the premises might be necessary the scheme has a very real advantage over renewal proceedings at court — as it is extremely difficult to arrange site visits to be attended by the judge.

The parties must consent to using the scheme. Once they do so the court will issue an order staying the claim and the parties make their joint application to the Law Society or the Dispute Resolution Service of the RICS. The application form is straightforward. As well as providing details of the parties and their representatives it is also necessary to outline the nature of the dispute, any specialist knowledge or experience which the expert/arbitrator will require and any potential conflicts of interest — the parties may name any individuals who they consider should not be appointed.

Arbitrations are carried out under the provisions of the Arbitration Act 1996, which oblige the arbitrator to act impartially. An arbitrator's award is enforceable by the courts. An expert determination is a procedure by which the parties to a dispute refer the dispute to an independent third party and agree to be contractually bound by the expert's decision (although parties often agree not to be bound by obvious or manifest errors, such as arithmetical mistakes, in the decision). In practice the chief difference between the two is that the arbitrator will reach a decision based largely upon the submissions of the parties. The expert, however, may take anything else which is relevant into account and may reach a decision based on his own knowledge and experience. The key area where the difference between the two will come into play is in those cases where the only areas of dispute are the rent payable under the new tenancy, and interim rent. Where the difference between the parties is not substantial it would be sensible and cost effective to allow an independent expert to make the determination. But in cases where the difference is substantial it would be more sensible for the parties to instruct their own experts to prepare reports and obtain comparable evidence and then ask the arbitrator to decide the issue.

Three obvious advantages of the PACT scheme, compared to court proceedings, are its speed, informality and flexibility. The parties are generally free to agree their own timetable and their own procedures. For example, they may decide that rather than prepare formal lists of documents (as required in court proceedings), that they will simply prepare a bundle of agreed core documents that both parties will work from. They might also decide, in consultation with the arbitrator or expert, whether a site visit is desirable. They may also decide to relax the stricter rules of evidence which apply in court cases to the use of comparable rental evidence. The parties will also be free to decide on how communications between themselves and the arbitrator or expert are to take place. They might agree that all communications are to be conducted by e-mail (a practice most county courts do not adhere to), although any communications from one party to the arbitrator or expert should always be copied to the other party. It is conceivable that relatively straightforward cases referred to the scheme could be decided in a few weeks.

The arbitration or expert determination often proceeds as follows. The parties agree a set of core documents and prepare a list of issues in dispute. The arbitrator or expert may convene a meeting to discuss and agree directions, or directions can be agreed in writing. The parties will then exchange, by or on an agreed date, the written evidence (including, where appropriate, comparables) they intend to rely upon. Next, the parties will exchange written representations on the written evidence submitted by the other. The parties and the arbitrator or independent expert will then agree whether there needs to be a formal hearing, although this is rare in the case of an expert determination. The arbitrator or independent third party may decide to send written questions to either or both parties, or he may invite the parties to serve written questions or requested for information upon the other party. He will then make and publish his award or determination.

10.16 The duties and role of the expert witness

Whatever the forum — court, mediation or the PACT scheme — it is likely that the parties will appoint an expert to advise them and prepare and present evidence on their behalf. In complicated opposed claims they may appoint more than one expert.

Expert evidence may only be called in court cases where the court has given permission (CPR 35.4). This ought to be limited to evidence reasonably required to resolve the claim. The expert's overriding duty is to the court and not the party which appointed him. The court will usually specify that the expert evidence is to be in the form of a written report meeting the following criteria. It must be addressed to the court; set out the expert's qualifications; give details of the material the expert has relied upon to reach his conclusions; where there is a range of opinions on matters set out in the report, summarise that range of opinions and give the expert's own opinion and his reasons for holding that opinion; and contain a summary of opinions reached. It must also contain a summary of all the instructions and material given to the expert that the expert has relied upon in reaching his opinions; often the expert will append a copy of his written instructions to his report and either confirm he has received no supplemental oral instructions, or state what those oral instructions were — the instructions given to the expert will not be legally privileged. It must also contain a statement that the expert has understood his duty to the court and has complied with it and contain a statement of truth signed by the expert.

The court considered the duties owed to it by expert witnesses in *National Justice Compania Naviera SA v Prudential Assurance Company Ltd* [1993] 2 EGLR 183 (usually referred to as *The Ikarian Reefer*). Those duties were summarised as follows. Expert evidence should be the independent product of the expert, uninfluenced as to its form or content by the exigencies of litigation. An expert witness should never assume the role of an advocate. An expert should not only set out the facts or assumptions relied upon to reach his opinion, but he should also set out those facts which might detract from his concluded opinion. The duty of the expert is to provide the court with an unbiased and objective opinion of matters within the expert's area of expertise. An expert should make it clear if a matter or question falls outside his area of expertise. If an expert believes his opinion is incomplete because he has insufficient material upon which to reach a final opinion, he should say so. If an expert changes his mind after his report has been exchanged the other party to the litigation, and if appropriate the court, should be informed promptly. If the report refers to plans or other documents they should be appended to the report.

Since the advent of the CPR, the court has extended the statement of expert's duties given in *The Ikarian Reefer*, following *Anglo Group plc* v *Winther Brown & Co Ltd* (2000) (72 Con LR 118). Here the judge was extremely critical of the expert evidence given by the claimant's experts, finding that none of them were truly independent, such that their evidence was unreliable. Accordingly the expert must also provide independent assistance to the court at all stages in the proceedings — this applies to any meetings between experts as much as to when the experts are giving evidence in person at court. The expert should co-operate with the expert or experts appointed by the other side to try and agree facts and narrow the areas in dispute. This means that experts should voluntarily attend "without prejudice" meetings with their opposite numbers and try, at the conclusion of such meetings to prepare a written record of areas where they have reached agreement and areas where they remain in dispute. The experts should bear in mind that that written record has the function of assisting the court and reducing the amount of time spent between the parties litigating the outstanding issues. Similar criticism of expert evidence was given by the court in *Clonard Developments Ltd* v *Humberts* [1997] EGCS 124, when the judge found that the expert evidence called by both sides was partial. These requirements are mirrored in the Guidance Notes which the RICS has issued to its members.

Of course experts sometimes find the transition from owing a contractual duty to a party who is paying their fees to an overriding duty to the court to be a difficult transition. This is even more the case where the expert has been appointed from an early stage and has already given his principal robust advice. The expert will be under pressure to adhere to that advice and not to concede any points. Experts should therefore be careful to follow carefully what the courts and the RICS have said. In *London & Leeds Estates Ltd* v *Paribas Ltd* [1995] EGLR 102, a disputed rent review arbitration was referred to the court. The tenant successfully argued that the landlord's surveyor should disclose the written expert evidence he gave in an earlier, unrelated rent review, as in that case the expert had reached conclusions inconsistent with the conclusions expressed in the subsequent rent review under appeal. Similarly, in 2001 the Court of Appeal upheld an earlier decision of the court which had struck out a claim when it discovered that an expert had criticised certain technical equipment which he had earlier effectively approved of in a different claim (*Hammersmith Hospitals* v *Troup Bywaters & Anders* [2001] EWCA Civ 793, CA).

10.17 Use of a single joint expert

Where the parties to a dispute wish to call expert evidence in respect of a particular issue the court has the power to order that evidence on that issue can be given by a single expert only (CPR 35.7).The overriding justification for this is that the appointment of a single joint expert is sometimes perceived to be cheaper and quicker than having each side appoint their own expert (*Daniels* v *Walker* [2000] 1 WLR 1382).

The test is whether the issue in question falls within a substantially established area of knowledge and where it is unnecessary for the court to sample a range of opinions. This might appear to be an attractive option to the parties in claims under the Act. Yet this procedure has been rarely implemented since the advent of the CPR, except in cases involving claims for personal injury. The attractiveness is tainted by several issues. It is almost always the case that the parties have already instructed experts before the proceedings have even begun; that parties tend to wish to appoint experts known to them. Even if a single joint expert is appointed the parties will usually wish to retain their own experts to advise on the terms of reference for appointing the new expert and to advise upon the actual report that he prepares. When account is also taken of the time and expense of appointing the new single joint expert, the anticipated costs savings often turn out to be illusory. If the parties cannot agree on the identity of the single joint expert, the court has to step in to make the appointment, usually by directing that the appropriate professional body will make the appointment, and this simply creates delay — there is sometimes disagreement between the parties as to the terms of reference for the single expert. While the single expert has the ability in such circumstances to ask the court for guidance under the CPR. this also creates delay.

In addition, either party or both parties may be concerned as to their position if the report of the single joint expert is unfavourable to them. Where a party is unhappy with the report of the single joint expert, the court has power to give permission for the disgruntled party to appoint another expert to challenge the report of the single joint expert (*Daniels* v *Walker* [2000] 1 WLR 1382, CA) but in practice the likelihood of the court giving permission is fairly remote. It must be proportionate for the court to allow yet another expert so, for example, the likelihood is that any request to challenge the report of a single joint expert would not be permitted in unopposed claims unless the only issue in dispute was the amount of rent and the rent in question was substantial. In all probability the disgruntled party would have to demonstrate that the evidence of the single joint expert is strikingly wrong. In *Cosgrove* v *Pattison* [2001] CP Rep 68, the court was persuaded to permit a second expert upon it being satisfied that there was a real risk that the single joint expert was biased.

10.18 Contents of expert witness statements in unopposed renewals

In unopposed claims where the only dispute is as to the new rent, the expert where usually seek to rely on evidence of rental levels being achieved on comparable properties. In cases where the expert was directly involved in negotiation or determination of the rent at the comparable property there will not be a difficulty in the expert giving evidence of that comparable.

However, in cases where the negotiation or determination of the comparable in question did not directly involve either the expert or the other party's expert the evidence is not regarded by the court as first-hand evidence. Instead it is regarded as hearsay evidence and special care must be taken in presenting it, failing which the court will regard that evidence as less reliable than direct first hand evidence. To avoid this happening the expert should ensure that the legal advisors seek to agree the comparable evidence with the other party. This is often done by obtaining a letter from a valuer who was involved in the negotiation or determination of the comparable rent, setting out all the relevant facts of the transaction, and serving notice upon the other party to admit the facts of the comparable transaction. This is a procedure permitted under the CPR. If the other party refuses to admit the facts, and they are later proven to be accurate at trial (usually by calling the valuer with direct first hand knowledge to give evidence at trial) the refusing party is at risk of having an order made against it to pay the costs incurred by the other party in proving the relevant facts. Alternatively the legal advisors may serve a hearsay notice under the CPR indicating the hearsay evidence on which that they intend to rely.

In those unopposed claims where the only issue is a change in the terms of the lease, expert evidence may be completely unnecessary. However, it is for the party seeking the change to justify it (see 12.4) and if the change is likely to have an adverse financial impact on one of the parties, or if the proposed change is likely to have an impact on the passing rent under the new lease, expert evidence will probably be necessary. Even in cases where the proposed change in the drafting of the new lease may not have a financial impact on either party, or may have a negligible impact on the new passing rent (for example the drafting of an authorised guarantee agreement), the evidence of an expert solicitor can be called. In cases where there is a dispute as to the inclusion of a landlord's option to determine the lease where the premises are likely to be re-developed by the landlord in the future, there may be the need for evidence to be called demonstrating the potential of the property for the type of work that would be involved if the landlord was opposing renewal on ground (f) (see 8.7).

10.19 Contents of expert witness statements in opposed renewals

This will depend upon the ground or grounds of opposition relied upon by the landlord.

Ground (a) — failure to repair

The landlord's building surveyor will probably refer to an earlier schedule of dilapidations, and give an indication of the cost of rectification. In a well-prepared case the landlord and the tenant will have set out their respective contentions in the form of a Scott Schedule. This is a document, in A3 size, which contains the relevant evidence in numbered columns. The columns will be individually headed, the normal headings and content of the columns being a description of the disrepair and where it is located; a reference to the clause in the lease it is contended by the landlord has been broken; the alleged cost of repair, including the contractor's preliminary costs and the professional fees the landlord expects to incur for having the work tendered and supervised; the tenant's comments upon the alleged disrepairs; and the tenant's view of the cost being claimed. The tenant also ought to set out its proposals for remedying the defects that it admits, with a likely timescale. One column is left blank for the trial judge to insert his comments upon individual items.

Both the landlord and the tenant's expert will wish to refer to the Scott Schedule in their statements. If the tenant has offered to undertake the work within a specified time the tenant's expert will comment upon the feasibility of that assertion.

Ground (b) — persistent arrears of rent

The only probable area where expert evidence could conceivably be of any benefit is in the financial area. The landlord may wish to call expert accountancy evidence as to the tenant's financial status, and to persuade the court that the tenant's accounts demonstrate that if it is granted a new lease it will continue to be unable to pay the rent timeously.

Ground (c) — other breaches

This is a wide-ranging ground of opposition and the nature of the expert evidence to be called (if any) will depend on the breach. If there has been a non-admitted breach of the user clause, for instance, the parties may wish to call the evidence of a town and country planner. If the allegation is that there has been an unlawful subletting the landlord may wish to call the evidence of an expert valuer showing that the landlord has suffered a diminution in the value of its reversion

Ground (d) — offer of suitable alternative accommodation

This is a ground where the landlord is seeking to prove that it has offered accommodation to the tenant which is comparable to the premises currently demised to the tenant (see 8.5). This question of comparability is often best dealt with by expert evidence. The expert may deal with the size, layout, location and condition of the alternative premises which the landlord has offered

Ground (e) — subletting of part

The evidence of an expert valuer will be required. The evidence will deal with the amount of rent that could reasonably be obtained on a separate letting of the holding; what amount of rent could reasonably be obtained on a separate letting of the remainder of the property comprised in the superior tenancy; what amount of rent could reasonably be obtained on a letting of the whole (see 8.6).

Ground (f) — demolition and construction

Here the landlord will have to satisfy the court, among other things, of the amount of work it intends to have carried out (see 8.7). The landlord may also need to demonstrate that the works in question cannot be carried out under the ambit of the terms in the present lease, how long the works will take and how much they will cost. The content of the expert evidence will depend on how far the landlord's intention has been developed. If planning permission has not been applied for or obtained the landlord may need to call the evidence of an expert town and country planner who will have to confirm that the proposed works would obtain planning permission.

The landlord will probably also need to call expert evidence from an architect, engineer or building surveyor to explain the proposals in detail and explain how substantial the works are going to be in practice. The evidence from such professionals may have to be more wide-ranging, depending of course on the facts of each case, but it is not inconceivable that expert evidence will be needed to demonstrate that the landlord has formed its intention for sound reasons, and the expert evidence may need to cover other relevant issues. The current condition of the premises and the building or buildings housing them may be relevant, particularly if the existing buildings are shabby or uneconomically laid out then it will be easier for the landlord to justify any scheme of redevelopment. The present value of the landlord's interest in the premises and the building or buildings housing them, compared with the value of the landlord's interest after the proposed works have actually taken place, will often also be relevant. Although the court need not investigate whether implementing the landlord's intention makes financial sense, it is none the less fairly obvious that if the landlord's proposed scheme is highly profitable to the landlord the court will be more readily able to accept that the landlord has the requisite intention under the Act.

10.20 Registration as pending land action

Prior to the making of a court application, the tenant's rights of renewal will normally amount to interests which override a disposition of the landlord's interest in the property, by virtue of the tenant being in occupation of the premises, and those rights will therefore be binding on a new landlord (Land Registration Act 2002, sections 11(4)(b), 12(4)(b), 29(2)(a)(ii) and 30(2)(a)(ii) and Schedule 1, para 2, and Schedule 3, para 2).

However, a tenant's application for a new lease under the Act is almost certainly a "pending land action" within the definition in the Land Charges Act 1972, and as such the tenant's rights under the court application are outside the protection given to overriding interests mentioned above (2002 Act, section 87(3)). The tenant's rights under his court application will therefore not bind a purchaser for valuable consideration of the landlord's interest in the property unless the court application is protected by an appropriate notice on the landlord's title register. An application for such a notice, called a unilateral notice, may be made to the appropriate office of the Land Registry under section 34 of the 2002 Act, using Land Registry form UN1.

A tenant is at risk of losing his right to a new tenancy if the landlord sells his reversion after the tenant has issued proceedings for a new tenancy but before the Land Registry has entered the notice of the court application on the landlord's registered title. However the tenant cannot apply to the Land Registry to enter that notice until the court proceedings have actually been issued and there will undoubtedly be a delay between the issue of the application by the court and the receipt by the Land Registry of the form UN1 from the tenant. To minimise this period of risk, it is possible for the tenant's solicitors, as soon as they know that the court has issued the court application, to make an immediate "outline application" to the Land Registry in respect of the intended registration of the unilateral notice. The outline application can, at times when the Land Registry provides these facilities, be made over the telephone or on-line, but only by solicitors who have the necessary credit or subscription facilities. If an outline application is made and the UN1 is actually received by the appropriate office of the Land Registry by noon on the fourth business day following the making of the outline application, the Land Registry will register the unilateral notice with effect from the time of the outline application (Land Registration Rules 2003, rule 54(9)).

Where the landlord's title is unregistered, the court application can be registered at the Land Charges Department of the Land Registry as a pending land action under the Land Charges Act 1972.

10.21 Striking out and dismissal for lack of protection

Where the tenant plainly does not have the protection of the Act because he is permanently out of occupation, any application to the court is liable to be struck out at the application of the landlord. However, the permanency of vacating turns on the facts and intention.

Casenote

Domer v *Gulf Oil (Great Britain) Ltd* (1975) 119 SJ 392: The tenant had vacated two months before making his application to the court. Held: The application was struck out as an abuse of the court.

10.22 Delays in proposed implementation of grounds (d), (e) or (f)

Under section 31(2), in cases where the landlord is opposing a new tenancy and the court is satisfied that he would have succeeded in proving an intention falling within grounds (d), (e) or (f) of section 30(1) of the Act — set out in Chapter 8 — but for being unable to prove that the ground would apply at the time of the termination of the current tenancy, the tenant will nevertheless be refused a new tenancy if the landlord can show that the ground will be satisfied at a later date, so long as it is not more than a year later than the expiry date specified in the original section 25 notice or section 26 request.

In that situation, section 31(2)(b) entitles the court, if requested by the tenant, to order the current tenancy to continue until that later date and then end with no new tenancy being granted.

10.23 Effect of court application on termination date

Where a section 25 notice or section 26 request is served and is followed by a court application, the termination date arising under the section 25 notice (being the termination date specified in it) or under the section 26 request (being the day before the date specified in it for the new tenancy to start) may be postponed by the operation of section 64.

The provisions of section 64 apply where (a) a notice to terminate a tenancy (which might be a section 25 notice or a section 27(2) notice) or request for a new tenancy (ie a section 26 request) has been given; and (b) the tenant has made an application to the court for a new tenancy; and (c) the termination date arising under the notice or request falls earlier than three months beyond the date on which the court application is "finally disposed of". The effect of section 64 is to postpone termination of the tenancy to three months after the court application is "finally disposed of".

The application is "finally disposed of" for the purpose of section 64(2) when

(a) the court application is withdrawn or discontinued or
(b) the county court or High Court makes an order and neither party appeals it to the Court of Appeal within the appropriate time limit, currently 14 days of the date of the court order or
(c) there is such an appeal, the Court of Appeal makes an order and neither party appeals to the House of Lords or applies to it for leave to appeal within the appropriate time limit, currently 28 days of the date of the Court of Appeal's order or
(d) such an appeal or application is made and the House of Lords refuses leave or hears the appeal and makes its decision (section 64(2); and see *Austin Reed Ltd* v *Royal Insurance Co Ltd (No 1)* [1956] 1 WLR 1339 for the position prior to the CPR and to the 2004 amendments).

Thus in cases where the tenant does not appeal against the first instance decision, the continuation tenancy ends three months and 14 days after the date of the court order. Accordingly, a tenant who has been seeking a new tenancy but changes his mind should have any court application withdrawn or discontinued at the earliest possible date.

The relationship between section 64 and section 27(2) is discussed at 4.7.

10.24 Negotiation, "without prejudice" and privilege

Negotiation is encouraged with a view to the parties reaching an agreed resolution to their dispute without having to resort to mediation, court hearings, or PACT. Inevitably the parties will make proposals or offers to each other in order to compromise areas of disagreement. Normally a party making such an offer will not want details of the offer to be disclosed to a judge or arbitrator during proceedings if the dispute is not resolved, since it might weaken his case or otherwise assist the other party. In order to protect such genuine offers and other concessions, made in an attempt to settle, from disclosure in proceedings, the party making the offer should put it forward expressly on a "without prejudice" basis. That phrase can be used both in oral discussions and in written communications. An offer to settle a dispute made on a "without prejudice" basis will generally be privileged and may not be referred to by either party in court proceedings relating to that dispute between those parties (see 10.12, 10.14, and 10.16). The "without prejudice" tag does not protect from disclosure anything other than offers put forward for the purpose of settling a dispute which, if not resolved, is to be the subject of proceedings before a court or arbitrator. However, in the present context such proceedings need not be imminent or inevitable; "without prejudice" can be used for an "opening shot" intended to start the negotiation process — see the rent review case of *South Shropshire District Council* v *Amos* [1986] 2 EGLR 194, which is assumed equally to apply to the negotiation of renewal terms under the Act. As for making offers without prejudice save as to costs, see 10.24.

There is a separate doctrine which requires a party to proceedings to disclose to the other party relevant documents under his possession or control. Only certain categories of documents are protected from this disclosure by the rules of privilege. Legal advice, whether relating to litigious business or not, is generally privileged. Communications between a solicitor and non-lawyers in connection with proceedings, such as an expert's report which is not put forward in evidence, are subject to "litigation privilege" (*Jackson* v *Marley Davenport Ltd* [2005] 1 EGLR 103, CA) However, documents which are prepared by non-lawyers for purposes other than the dispute (eg valuations for mortgage purposes) may not be protected from disclosure if they contain information relevant to the dispute, such as assessments of rental value. The parties and their advisers should be cautious about allowing such documents to be prepared when lease renewal proceedings are current or anticipated. Further discussion of the detailed rules of privilege is outside the scope of this book.

The expression "subject to contract" must not be confused with "without prejudice". Making an offer on a "subject to contract" basis means that acceptance of the offer will not create a legally binding agreement. The formalities for reaching a binding agreement on the terms for the new tenancy are discussed at 13.1 and 13.2.

10.25 Costs of proceedings and new leases

It is common for there to be no order as to costs when the court has to determine the terms of a new tenancy under the 1954 Act; each party must generally bear its own costs. This is for two main reasons. First, there may be several different matters in issue, some inter-connected and others not, and a party may succeed on some but not on others. Second, the risk of becoming liable for the landlord's costs might discourage some tenants from exercising their rights under the Act. Consistently with this approach, the courts will not even order the new lease to contain a covenant by the tenant to pay the landlord's costs for preparing the new lease (*Cairnplace Ltd* v *CBL (Property Investment) Ltd* [1984] 1 EGLR 69, CA). This approach will prevail even if the tenant's existing lease contains a covenant by him to pay the costs incurred by the landlord in any renewal procedures, since such a covenant will be invalid under section 38(1) because in effect it imposes a penalty on the tenant if he applies to the court (*Stevenson & Rush (Holdings) Ltd* v *Langdon* (1979) 1 EGLR 72, CA).

However, in some cases a party might make an offer of settlement in a manner which puts the other party at risk of having to pay some of the costs incurred by the offering party if he does not accept the offer. The concept is that if such an offer is made and rejected and subsequently the matter is determined in proceedings such that the party who received the offer is no better off than if he had accepted the offer, it is reasonable for him to pay the offeror's costs incurred after the expiry of a reasonable time given for acceptance of the offer. In practical terms this tactic is most appropriate in those lease renewal cases where there is only one matter in dispute, such as the amount of the new rent, so that the outcome will be clear cut. If his offer is not accepted, the offeror will normally want the details of the offer to be withheld from the court until the substantive issue in dispute has been decided, since disclosure of the terms of his offer might adversely affect his ability to argue for a more advantageous outcome in the proceedings.

To make an offer on this basis, certain formalities must be observed. Where the dispute will go to court if not resolved, it may be possible to make the offer in accordance with the detailed requirements of Part 36 of the CPR and the Practice Direction — Offers to Settle and Payments into Court. In such cases the consequences in respect of costs will be as laid down in Part 36, broadly in line with the concept outlined above. More details of Part 36 of the CPR are outside the scope of this book.

Where it is not appropriate to comply with CPR Part 36, or where the dispute will go to arbitration instead of the court, there is the alternative possibility of putting forward the offer as a "Calderbank" letter (so called after the matrimonial case of *Calderbank* v *Calderbank* [1975] 3 All ER 333, CA). This is a letter making an offer on a without prejudice basis but stating that the offer, if not accepted, may be brought to the attention of the court (or, if applicable, an arbitrator), after the matter in dispute is determined, in relation to costs. A common short formulation is "without prejudice save as to costs". The effect is that the judge (or arbitrator) can take the offer into account when exercising his discretion on costs, but the details of the offer are protected from disclosure to him earlier in the proceedings and cannot affect his determination of the actual issue in dispute. A *Calderbank* offer must cover all the matters in dispute, including liability for costs, and must be unambiguous so as to be capable of acceptance by the offeree — it must not be made "subject to contract" or subject to any further approval from the offeror. It must allow the offeree a reasonable period in which to consider and accept the offer. Unlike Part 36 offers, the costs consequences of a successful *Calderbank* tactic remain completely at the discretion of the court and the cost consequences that would apply to a similar outcome under a similar offer made under Part 36 will not automatically follow (CPR Rule 36.1(2)). The existence of the *Calderbank* letter therefore cannot decisively govern the outcome as to costs, but nevertheless it should influence the court when exercising its discretion on costs.

In arbitration cases, a copy of an unaccepted *Calderbank* offer letter may be given to the arbitrator during the proceedings as a "sealed bid" not to be opened until he has made his award on the substantive issues and is dealing with the issue of costs. Guidance for arbitrators was given by Donaldson J in *Tramountanta* v *Atlantic Shipping Co SA* [1978] 2 All ER 870 in relation to the effect of *Calderbank* offers on costs, but today judges and arbitrators tend to exercise their discretion and powers as to the award of costs in a manner that is designed to reward the true winner in a case, if there is one. Making a genuine *Calderbank* offer during the course of a dispute may well have the desired effect on the offeree because of the possible costs consequences of rejecting the offer, but at the end of the day a judge or arbitrator will be obliged to do no more than have some regard to any such offer, and in many cases under the Act there will be no order as to costs for the reasons set out above.

Interim Rent

11

11.1 Interim rent applications

Following the 2004 amendments, and while renewals continue to flow from section 25 notices and section 26 requests served before 1 June 2004, there are two parallel regimes operating in respect of interim rent. For cases which are governed by the Act prior to the 2004 amendments (see 1.2), only the landlord can apply for an order for interim rent, under the unamended version of section 24A and the old rules abut ascertaining the interim rent apply. For cases to which the amended Act applies, either the landlord or the tenant may apply to the court for an order for interim rent, under the new version of section 24A, and the interim rent is assessed under new rules.

In cases to which the new rules apply, to avoid both parties making a court application for interim rent, once one party applies for it, the other may not do so unless the first party withdraws his application (new section 24A(2)). The provisions unfortunately do not address the question as to how one party will actually know if the other has made an interim rent application until he has been served with a copy of it.

Also under the new rules, neither party may apply for an interim rent order once six months has elapsed after the end of the continuation tenancy to which the interim rent would relate (new section 24A(3)). However, unlike the provisions of section 29 (relating to applications to the court regarding the grant of a new tenancy), there is no provision here which prohibits a party from withdrawing his application without the consent of the other party. Accordingly, if an application for interim rent is made by one party but the other party believes that he will benefit from it, that other party should ensure that the application is brought before the court before the six month time-limit expires, in case the applying party were to withdraw his application at a time when it is too late for the other to make his own application.

11.2 Start date for interim rent

The interim rent, once agreed by the parties or fixed by the court, will run from a particular date. In relation to cases which are governed by the Act prior to the 2004 amendments (see 1.2), it will run from the termination date arising under the section 25 notice or section 26 request or, if later, the date when the landlord applies to the court for the interim rent order.

In relation to cases covered by the 2004 amendments, the interim rent will instead be payable as from the earliest date which could have been specified as the termination date in the landlord's actual section 25 notice or, in the case where the tenant gave a section 26 request, the earliest date which could have been specified in that request as the date for the new tenancy to begin (section 24B(2),(3)). As for identifying those earliest possible dates, see 6.4 and 7.3.

In either case the court may, upon hearing the application, order an interim rent to be paid. The word "may" appears in both the new and old versions of section 24A(1). This means that the court is not obliged to order interim rent, but can refuse to make an order for interim rent if the circumstances justify, although this is expected to be rare. One instance might be where the landlord has neglected to serve a section 25 notice on a subtenant after becoming the competent landlord (see 9.2), thus delaying the ability of the tenant to obtain an increased rent from the subtenant.

If the court does not order interim rent to be paid and the parties do not reach an agreement that interim rent is to be paid, the rent payable under the current tenancy will continue to be payable throughout the continuation tenancy.

11.3 The two alternative bases of assessing interim rent

There are two broad bases for assessing interim rent. The first applies only to new cases covered by the 2004 amendments (see 1.2) where the tenant occupies, and renews his tenancy of, the whole of his premises without opposition by the landlord. The second applies to cases covered by the law prior to the 2004 amendments and also to those new cases where the landlord opposed renewal or where there are subtenants or vacant parts of the premises.

11.4 Interim rent at an unopposed renewal of whole with no subtenants

For cases covered by the 2004 amendments (see 1.2), but subject as mentioned below, the interim rent will be fixed under the new section 24C at the same amount as the rent to be payable under the new tenancy (determined in accordance with section 34 — see 12.6) in those cases where all the following facts occur — the landlord gives the tenant a section 25 notice or the tenant gives the landlord a s26 request for a new tenancy; the landlord does not state his opposition (either in the section 25 notice or by giving a section 26(6) counter-notice) to the grant of a new tenancy; at the time of the section 25 notice or the section 26 request the tenant occupies the whole of the premises comprised in his tenancy (with no vacant parts and no sub-tenants); and the tenant is actually granted a new tenancy of the whole of those premises either by agreement or by court order (section 24C(1),(2)). The theory is that, in these particular circumstances, there is usually no reason why the interim rent should not be the same as the rent that will be payable under the new tenancy.

Nevertheless, even where all the above facts apply, either party may apply to the court to order an interim rent which is different from the rent fixed under s.34 for the new tenancy, in three alternative situations.

The first situation is where there is a significant change in market rents (either an increase or a decrease) between the start date for the interim rent and the date at which the rent for the new tenancy is assessed. The party seeking to argue this must satisfy the court that the amount of rent fixed under section 34 for the new tenancy — which will reflect market rental values at the date of the court hearing with any appropriate adjustment for the three or four month delay until the new tenancy will start — is substantially different from the amount that would have been fixed under section 34 if the new tenancy had started on the date from which the interim rent runs (section 24C(3)(a),(4)). If the court is satisfied as to that, the interim rent will be set at that alternative amount (section 24C(5)).

The second situation is where the tenancy terms are being substantially changed from those of the current tenancy, for example a change from internal or limited repairing terms to full repairing terms or vice versa, and so the interim rent (which relates to a period in which the old terms continue to apply) should reflect the old terms and not the new terms. The party arguing this must show to the court's satisfaction that the rent fixed under section 34 for the new tenancy — which will reflect the other terms on which the new tenancy is to be granted — would have been substantially different if the other terms of the tenancy were the same as in the current tenancy (section 24C(3)(b)). If the court is satisfied as to that, the interim rent is instead to be a rent which it is reasonable for the tenant to pay in respect of the interim rent period (section 24C(6)).

The third situation is where both the first and second sets of circumstances apply — substantial changes in the market rent coupled with substantial differences in rental values due to changing the terms of the tenancy. In this mixed situation, again the interim rent is to be that which it is reasonable for the tenant to pay in respect of the interim rent period (section 23C(6)).

In both the second and third situations, it is provided that in determining the rent which it is reasonable for the tenant to pay in respect of the interim rent period under section 23C(6), the court shall have regard to the rent payable under the current tenancy (before the start of interim rent) and the rent payable to the tenant under any subtenancy of part of the property (that is, if the tenant has granted a subtenancy of part since the date of the section 25 notice or section 26 request) and that otherwise the rent is to be that which would fixed on the formula in section 34(1) and (2) (see 12.6) on the basis of a fixed term lease starting on the date from which the interim rent starts and of a length equal to the fixed term to be granted by the new tenancy (section 24C(7)). As for the impact on rent of the direction to have regard to the current rent, see the discussion about "cushioning" below. Section 24C(7) is silent as to the terms of the tenancy that are to be reflected in the rent under this formula, but presumably these will be the terms of the current tenancy, since they continue to apply during the continuation period (see 4.2); if the parties agree, or the court determines, that the new tenancy is to be on different terms, those new terms will not take effect until the new tenancy commences.

If the court orders the grant of a new tenancy and orders interim rent to be paid but the new tenancy is not granted, either following the tenant obtaining a revocation of the court order or by agreement between the parties, either party can apply to the court to determine a new interim rent (section 24D(3)). Thus where interim rent is fixed under section 24C in the belief that the new tenancy will be taken up but it is not, a new interim rent can be fixed under section 24D as discussed below.

11.5 Interim rent in other situations

Where one or more of the facts required for the application of section 24C do not apply, the interim rent will be governed by section 24D instead. Accordingly, section 24D applies where the landlord states his opposition (either in the section 25 notice or by giving a s.26(6) counter-notice) to the grant of a new tenancy, even if he subsequently withdraws that opposition; or at the time of the section 25 notice or the s.26 request the tenant occupies only part of the premises comprised in his tenancy; or the tenant does not take up the grant of a new tenancy; or the tenant accepts the grant of a new tenancy of only part and not the whole of the premises comprised in his current tenancy.

If the court makes an order for interim rent under section 24D, it is provided that the amount of that rent is to be that which it would be reasonable for the tenant to pay while the tenancy continues by virtue of section 24 and in determining that amount the court shall have regard to the rent payable under the terms of the current tenancy and the rent payable to the tenant under any subtenancy of part of the property and that otherwise the rent is to be that which would fixed on the formula in section 34(1) and (2) on the basis of the grant of a new tenancy from year to year of the whole of the property comprised in the current tenancy (section 24D(1),(2)).

The formula in section 24D is based on that in the old section 24A, but is subject to the new rules in section 24A allowing either party to apply for the interim rent and the new rules in section 24B as to the start date for interim rent. Apart from those new rules, the similarities between the formulae in the new section 24D and in the old section 24A mean that guidance may still be derived from cases decided on the old section 24A formula. The following consideration is based on those cases.

The rent valuation date for the interim rent under this formula is the date from which the interim rent will run (*Janes (Gowns) Ltd* v *Harlow Development Corporation* [1980] 1 ELR 52). The "tenancy from year to year" basis of valuation has in the past frequently produced an interim rent of 10% to 20% below the market rent that will be payable under the new tenancy, but this depends on various valuation factors and is a matter of valuation evidence. During depressed market conditions, there could be an argument that a tenancy from year to year produces an interim rent of 10% or so above the market rent payable for a fixed term of years, reflecting the tenant's ability to renegotiate yearly in a falling market. These are matters for valuation evidence.

The direction to "have regard to" the rent under the current tenancy is designed to enable the judge, if he thinks it appropriate, to fix the interim rent at a level which "cushions" the tenant against a drastic increase in rent for the continuation period compared with the rent that had been payable under the existing tenancy (*English Exports (London) Ltd* v *Eldonwall Ltd* (1972) 225 EG 255, ChD; *Fawke* v *Viscount Chelsea* [1979] 1 EGLR 89). The question whether to give the tenant a cushioning discount off the yearly rental, and if so how much, is a question for the judge to determine in his discretion, on the facts of each case; this factor is not a matter for valuation evidence. The cushioning discount is not mandatory (*Conway* v *Arthur* [1988] 2 EGLR 113, CA). In some cases no cushioning discount will be justified, for example perhaps where the past rent has been considerably below market level (*Department of the Environment* v *Allied Freehold Property Trust Ltd* [1992] 2 EGLR 100, Mayors and City of London Court). This is a usually matter for the judge in his discretion and his judgment will not be overriden by the Court of Appeal unless, exceptionally, it considers that the judge has ignored a material consideration (*French* v *Commercial Union Assurance Co plc* [1993] 1 EGLR 113, CA) .

Casenotes

English Exports (London) Ltd v *Eldonwall Ltd* (1972) 225 EG 255, ChD: The old rent was £7,655 pa. The new tenancy rent was to be £16,000 pa. Held: The interim rent is to be £15,000 pa.

Fawke v *Viscount Chelsea* [1979] 1 EGLR 89, CA: Major repairs to the premises were required because of the landlord's failure to perform his covenant for external repairs. Held: The court should fix a reduced interim rent until the landlord completed the work, to increase at that time to the appropriate level.

Janes (Gowns) Ltd v *Harlow Development Corporation* [1980] 1 EGLR 52, ChD: The old rent was £2,100 pa. The new rent was to be £12,500 pa. Held: Interim rent £9,200 pa.

UDS Tailoring v *BL Holdings Ltd* (1981) 261 EG 49: The interim rent was fixed at 10% below the new rent.

Ratners (Jewellers) Ltd v *Lemnoll Ltd* [1980] 2 EGLR 65: The old rent was £11,250. The new rent was to be £24,250. Held: The interim rent should reflect a 15% discount for the yearly tenancy basis, producing £19,582, and should be further reduced to £17,500 for the cushioning.

Charles Follett Ltd v *Cabtell Investments Ltd* [1987] 2 EGLR 88: The old rent was £13,500 pa. The new tenancy rent was to be £106,000 pa. A yearly tenancy rent would have been £80,000 pa. Held: Interim rent £40,000 pa (but since the landlord had originally sought only £25,000 pa as the new rent, the tenant need only pay that amount for the period of the original negotiations). (NB: In the French case cited at 12.6, the Court of Appeal observed that the 50% reduction from £80,000 to £40,000 in *Cabtell* was exceptional.)

Woodbridge v *Westminster Press Ltd* [1987] 2 EGLR 97, ChD: The old rent was £5,000 pa. The new rent was to be £29,000 pa for the nine month new tenancy; it would have been £52,000 pa for a seven year term. The yearly tenancy figure was £30,500 pa. Held: No cushion was justified.

Conway v *Arthur* [1988] 2 EGLR 113, CA: The tenancy ended in December 1984. The current rent was £1,250 pa fixed at a review in 1979. The terms of the new tenancy were not determined until a county court hearing in January 1988, at which the judge fixed a new rent of £4,981.50 pa and an interim rent backdated to December 1984 of £4,234 pa. The judge did not give reasons for this level of interim rent, the landlord's surveyor had not given evidence on the issue and the tenant's surveyor had suggested £2,500 pa. The Court of Appeal held: (1) there was no requirement for cushioning in every case, but each case had to be decided on its own facts; (2) since the landlord's surveyor had not given evidence on the interim rent, there was no material on which the judge could depart from the tenant's surveyor's evidence, and the case was remitted back to the county court judge for a fresh hearing on the issue.

Department of the Environment v *Allied Freehold Property Trust Ltd* [1992] 2 EGLR 100, HH Judge Hague QC, Mayor's and City of London Court: The tenant had the benefit of a very low rent during inflationary times. Held: There was no need to perpetuate what inflation and the passage of time had turned into an injustice to the landlord; no cushioning discount.

Simonite v *Sheffield City Council* [1992] 1 EGLR 105, ChD: Interim rents were fixed at 25% below the new rent to reflect an annual tenancy. No cushioning discount was mentioned in the judgment.

11.6 SDLT on interim rent

The possible liability of the tenant to pay SDLT in respect of the passing rent during his continuation tenancy has been explained at 4.10 and initially applies only where the tenant's lease was granted on or after 1 December 2003 so as to fall within the SDLT regime. However, whether or not that lease was subject to old stamp duty or SDLT, the tenant might have a separate liability to pay SDLT if the amount agreed or determined as the interim rent is higher than the passing rent. In such cases the increase is normally likely to be treated as a variation of the rent under the tenant's lease and may give rise to SDLT being payable under schedule 17A para 13 of the Finance Act 2003, in which case a return must be submitted and the SDLT paid within 30 days of the making of the court order or the legally binding agreement as to the interim rent.

There might be a further SDLT consequence if the agreement or determination of the amount of interim rent is part of an arrangement which includes the grant of the new lease. It is often the case that interim rent is agreed by the parties, or fixed by the court, at the same time as they agree or determine the terms of the new lease. Indeed, the interim rent will in some cases be automatically set at the same amount as the rent under the new lease (see 11.4). Where there is this linkage between the interim rent and the grant of the new lease, it seems that varying the rent for the interim period to the level of the interim rent and granting the new lease may be treated as "linked transactions" within schedule 17A para 5 of the Finance Act 2003. If so, SDLT might be payable on them as a combined transaction by calculating it, under the formula mentioned at 4.10, as if the new lease started on the date from which the interim rent increase runs and then continued with the actual term of the new lease, with the rent for that additional notional initial period being taken as the amount by which the interim rent exceeds the previous passing rent.

However, it is possible that even more SDLT may be payable if the term of the new lease is actually backdated to include some or all of the interim rent period (see 12.3 and 12.12).

The New Tenancy

12

12.1 Premises and rights to be granted

The tenant's right to apply for a new tenancy relates only to the "holding" — those parts of the premises let to him which he is actually occupying (section 23(3), section 32(1)). Where the holding is only part of the premises comprised in his existing tenancy (for example, where other parts are vacant or sublet), the landlord has the right at his discretion to insist that any new tenancy to the tenant must comprise the whole of that original demise and not just the holding (section 32(2)). If the tenant is not willing in that situation to take a new tenancy of the whole premises, he must give up his right of renewal.

Thus where the tenant does not occupy the whole premises let to him, the landlord can choose between granting a new tenancy of just the holding and granting a tenancy of the whole of those premises. Subject to one exception, the Act does not provide for the tenant to take a new tenancy of something in between, less than the original demise but more than the holding — it is one or the other. The exception is where section 31A applies because the landlord is opposing renewal of ground (f) but intends to redevelop only part of the premises and the tenant agrees to take a new tenancy of just part of the holding (see 8.8).

Rights enjoyed by the tenant under the terms of the existing tenancy in connection with the holding are to be included in the new tenancy except as otherwise agreed by the landlord and the tenant or determined by the court: (section 32(3)). The Court of Appeal has said that the courts have no power under either that section or section 35 (discussed below in connection with the other terms of the tenancy) to grant the tenant new rights which he does not have under his existing tenancy (*G Orlik (Meat Products) Ltd* v *Hastings & Thanet Building Society* [1975] 1 EGLR 70, CA). However, it may be that this only applies where the new tenancy comprises the whole of the premises let to the tenant by the existing tenancy; where only the holding (not constituting all the original premises) is to be comprised in the new tenancy, the court may consider itself able to insert necessary rights over the remainder of the premises (see *Re 5 Panton Street, Haymarket* (1960) 175 EG 49, ChD).

The "rights" to be included under section 32(3) do not include rights granted as terms of the existing tenancy but which are of a personal nature unconnected with the occupation of the holding. However, it may be that such terms can be repeated in the new tenancy by virtue of section 35 (see below).

Casenotes

Re No 1 Albemarle Street London W1 [1959] 1 All ER 250: The lease of second floor offices contained a separate clause giving the tenant the right to display advertisements of its products on the outside of the building, unconnected with its offices. The landlord objected to repeating this right in the new tenancy. Held: This right was a personal right outside the scope of section 32(3). However, it was within the scope of section 35 and should be repeated by way of licence under the new tenancy.

Re 5 Panton Street, Haymarket (1960) 175 EG 49, ChD: The existing tenancy comprised a five-storey building but the new tenancy was to comprise only the tenant's holding, being restaurant premises on the basement, ground and first floors. The tenant then needed rights over the remainder for access and ventilation. Held: The new tenancy would contain rights for ventilation and access over the upper part of the building, and a landlord's covenant to maintain the shared entrance, stairs, passages and landing, subject to the tenant paying a maintenance contribution.

G Orlik (Meat Products) Ltd v Hastings & Thanet Building Society [1975] 1 EGLR 70, CA: The tenant had parked vehicles on an adjacent yard owned and occupied by his landlord. The use of the yard for parking was not specifically granted by the tenant's lease and the landlord objected to including that right in the new tenancy. Held: Such a right was outside section 32(3) since, in the absence of rectification, this arrangement was not a right enjoyed under the existing tenancy. Further, the court has no power under section 32(3) (or under section 35) to deal with this issue by enlarging the tenant's holding to include additional areas in the demise.

12.2 Rights to be reserved

The Act is silent as to rights over the premises which are excepted and reserved under the terms of the existing tenancy, but it is generally considered that, without good reasons to the contrary, these are to be repeated in the new tenancy. The court will however be reluctant to insert new exceptions and reservations (*Heath* v *Drown* [1973] AC 498, HL).

12.3 Duration of the new tenancy

Section 33 of the Act requires the court to determine the duration of the new tenancy "to be reasonable in all the circumstances". However, there is a maximum duration for a new fixed term tenancy which the court can order unless both parties agree otherwise. For cases governed by the Act prior to the 2004 amendments (see 1.2) the maximum term is 14 years. For cases governed by the 2004 amendments it is 15 years.

The new tenancy is to begin on the coming to an end of the current tenancy. Where a court application has been made (see Chapter 10) this end date for the current tenancy is governed by section 64 (discussed at 10.23) and is usually three months and two weeks after the date of the court order for the new tenancy. In many cases it is not the same date as the expiry date of the contractual term or the expiry date of the section 25 notice or section 26 request. However the parties may voluntarily complete the new lease earlier, by agreement. Since the date on which the existing tenancy will end depends upon the date of the court hearing and whether or not there is an appeal, the exact date for the new tenancy to commence will not be known for certain at the date of a first instance court hearing. For this reason the landlord's proposals for the new term, and the court order, should specify a fixed expiry date for the new tenancy, rather than a number of years for the duration of the new term (*Warwick & Warwick (Philately) Ltd* v *Shell UK Ltd* [1981] 1 EGLR 51, CA). This ensures that any delays due to the court procedure will not become cumulative.

Further, there may be adverse SDLT consequences for the tenant if the start of the term were backdated to a date before the new lease is granted (see 12.12).

If the parties cannot agree the length of the new term, the court will order such term as it considers reasonable in all the circumstances. The court is usually likely to order a new term of such duration as is requested by the tenant (provided it is no longer than 15 years), particularly where the tenant has occupied the premises for a long time. However, in some cases the landlord's requirements will be taken into account, especially where he asks that the term should be coterminous with the leases of adjacent premises.

If a tenant requests only a short new tenancy, the court is likely to accede to this request provided the term is not so short as to prejudice the landlord; if a very short period is sought by the tenant, it should be sufficiently long to enable the landlord to have a reasonable opportunity to find a new tenant.

The question of inserting a break clause in the new tenancy, for example entitling the landlord to determine the new term for the purpose of redevelopment, is normally considered under section 35 (see 12.4).

In some cases the landlord may wish to oppose a new tenancy on ground (e), (f) or (g) (see Chapter 8) but cannot succeed because he cannot demonstrate that his intended works or use can be implemented sufficiently soon after the end of the current tenancy, or he may not yet have owned the property for the full five years required to succeed on ground (g). In such cases the landlord may be able to persuade the court to order the duration of the new tenancy to be short or determinable on notice, enabling the landlord to make a fresh attempt to obtain possession in the near future on the narrowly missed ground of opposition. However, to be fair to the tenant, the term should not be so short that the tenant has insufficient time to relocate his business.

Casenotes

Re Sunlight House Manchester (1959) 173 EG 311: An office block had been let to the Ministry of Works for 27 years on a series of leases, the last being for ten years. The Ministry was due to relocate very soon and sought only a six months' new tenancy, but the landlord asked the court to order a new term of three years. Held: The landlord should have a reasonable time to relet the premises. The new tenancy should be for ten months.

Turone v *Howard de Walden Estates Ltd* [1983] 2 EGLR 65, CA: The old lease was for ten years. The landlord served a section 25 notice not opposing a new tenancy, but proposed that it should be for seven years so as to be coterminous with the leases of adjacent premises of which he was also the landlord. The tenant asked for a ten year term. Held: The landlord's proposal was accepted. (Due to delays under section 64 and the county court judge making an inappropriate order, the final order was for the new tenancy to expire slightly later.)

CBS (UK) Ltd v *London Scottish Properties Ltd* [1985] 2 EGLR 125: The tenant of a large warehouse wanted a short new tenancy of just 12 months. The landlord sought an order for a new 14 year lease. Held: A short tenancy should be granted as requested by the tenant; it would be sufficiently long to give the landlord time to relet the premises.

Becker v *Hill Street Properties Ltd* [1990] EGCS 31: The tenant, a dentist, sought a new lease to end in just over four years, when he intended to retire. The landlord wanted him to vacate sooner and asked the court to order either a shorter term or the insertion of a landlord's break clause. Held: The tenant should be granted the term he requested, with no break clause.

Upsons Ltd v *E Robins Ltd* [1956] QB 131: The landlord's attempted opposition to a new tenancy under ground (g) infringed the five year rule under section 30(2) by just two months. Held: The words "all the circumstances" in section 33 were wide enough to take account of that situation and the new tenancy should be for only 12 months, at which time the landlord could make a fresh attempt to obtain possession on ground (g).

Reohorn v *Barry Corporation* [1956] 1 WLR 845, CA: The landlord's attempted opposition to a new tenancy on ground (f) failed because the landlord's negotiations with a developer for a major redevelopment were in an early stage, only outline plans had been prepared, and the negotiations and detailed planning were likely to take a long time. Held: The new tenancy should be terminable on six months' notice given at any time.

London & Provincial Millinery Stores Ltd v *Barclays Bank Ltd* [1962] 1 WLR 510: The landlord did not oppose the tenant's section 26 request for a new tenancy. The premises were in a block, the remainder of which was dilapidated and unoccupied by the time the case came to court nearly five years later. The landlord had contracted to sell the block to a developer for redevelopment. The tenant sought and was granted a new tenancy for a term of nine years. Held on appeal: All the circumstances should be taken into account, particularly the proposed redevelopment and the fact that the tenant had already enjoyed a continuation tenancy of over four years. The new tenancy to be for 12 months from the date of the Court of Appeal's judgment, just to give the tenant an opportunity to relocate.

12.4 Terms of the new tenancy other than duration and rent

Section 35(1) provides that, so far as the parties cannot agree, the terms of the new tenancy other than duration and rent are to be determined by the court and in doing so "the court shall have regard to the terms of the current tenancy and to all relevant circumstances". By section 35(2), these circumstances are to include the application of the Landlord and Tenant (Covenants) Act 1995, discussed below.

Unlike section 34 (rent — see 12.6), this section makes no express reference to what is happening in the open market, but starts by referring to the terms of the existing tenancy. As a result, neither party can insist that the new tenancy is to be on terms that would be likely to be negotiated on a new letting. To the contrary, the starting point is the existing tenancy and the burden is on a party seeking to depart from its terms to justify his proposed changes and to show that they are reasonable.

Each proposed departure from the terms of the existing tenancy must be dealt with on its own merits, although subject to the overriding question whether the proposed departure can be justified on the grounds of "essential fairness" between the landlord and the tenant (*O'May* v *City of London Real Property Co Ltd* [1982] 1 EGLR 76, HL).

In the absence of strong cogent evidence justifying a change, the court will normally reject a departure which prejudices the security of tenure of the tenant in his business, because that is the very thing which the 1954 Act is designed to prevent (*Gold* v *Brighton Corporation* [1956] 3 All ER 442).

The court will not allow a departure from the terms of the current lease (eg relaxing the restrictions on alienation or change of use) if it is sought by the landlord essentially for the purpose of increasing the rent (*Charles Clements (London) Ltd* v *Rank City Wall Ltd* [1978] 1 EGLR 47). Conversely if the current lease contains a wider user covenant than the tenant actually requires, the court will not, at the tenant's request, narrow the user covenant simply because this would result in a lower rent (*Aldwych Club Ltd* v *Copthall Property Ltd* (1962) 185 EG 219).

Nevertheless, modernisation of the text and layout of the lease is not precluded. In the *O'May* case, the House of Lords stated that section 35 was not to have the effect of petrifying or ossifying the terms of the lease. If the terms of the current tenancy are obsolete or deficient, the court may consider this an adequate reason for a change, although any change which substantially alters the rights and obligations of the parties must be considered under the above rules. Equally, the inclusion of a reasonable provision which is for the benefit of one party and will not disadvantage the other party if he acts properly, may be accepted, for example, a provision for interest to be paid at a reasonable rate on late payment of money due under the lease, or for payment of value added tax on rent.

In some instances the law may have changed since the existing lease was granted and the new tenancy will have to reflect the current law. Section 35(2) expressly mentions the Landlord and Tenant (Covenants) Act 1995. If the existing lease was granted before 1 January 1996, when the 1995 Act came into force, the tenant may have had ongoing liability throughout the term by virtue of the common law rule of "privity of contract". Nevertheless a new tenancy cannot be granted on that basis since section 25 of the 1995 Act invalidates any provisions which run contrary to that Act, which provides that a tenant will not generally have ongoing liability after a permitted assignment. The landlord in that situation may argue that, for the new tenancy, the assignment clause should be tightened up to impose financial tests for acceptable assignees and to require an assigning tenant to enter into an authorised guarantee agreement under section 16 of that Act. The court is however likely to be reluctant to impose greater restrictions on assignment than those contained in the existing lease and is likely to qualify any new provision for the landlord to require an authorised guarantee agreement so that he can only require it where it is reasonable to do so (*Wallis Fashion Group Ltd* v *CGU Life Assurance Ltd* [2000] 2 EGLR 49).

If the terms of the existing tenancy include covenants by a guarantor, there is no certainty that this will be repeated in the new lease. Further, the court has no power to order a guarantor to join in the new tenancy; the Act contains no provisions for existing guarantors to be a party to the negotiations or court proceedings and the parties are confined to the landlord and the tenant. Nevertheless, section 35 has been construed as allowing the court, if it so decides, to order that the terms of the new tenancy shall include covenants by a suitable guarantor, so that the tenant would have to forgo the renewal if he was unable or unwilling to procure the joining of the proposed guarantor (*Cairnplace Ltd* v *CBL (Property Investment) Co Ltd* [1984] 1 EGLR 69).

It seems that section 35 entitles the court to order the inclusion in the new tenancy of rights which are included in the terms of the existing tenancy even if they are of a personal nature unconnected with the occupation of the holding and therefore are outside section 32(3), discussed above. Where only the holding (not constituting all the original premises) is to be comprised in the new tenancy, the court may be able to insert necessary terms relating to the remainder of the premises (see the earlier discussion).

In some cases the landlord may propose to introduce a redevelopment break as a term of the new tenancy. It is well established that, having regard to the fact that redevelopment is a ground for opposing a new tenancy under section 30(1) of the Act, it is not the purpose of the Act to confer security of tenure on the tenant at the expense of denying the landlord the right to redevelop. The court will normally allow a redevelopment break clause to be inserted where the landlord can show that redevelopment of the property during the new term is a possibility (*Adams* v *Green* (1978) 2 EGLR 46). The landlord does not have to show that the necessary preliminaries for redevelopment such as planning permission are already in place, since such issues would be considered under ground (f) (see 8.7) if and when the break clause was exercised and a further new tenancy opposed on ground (f). However, the clause must be on terms which are reasonable to the tenant as well as to the landlord; the court will fix the earliest date on which the break clause will be exercisable by balancing the interests of landlord and tenant, as it does in the case of the other proposed changes to the lease terms (*Amika Motors Ltd* v *Colebrook (Holdings) Ltd* [1981] 2 EGLR 62, CA). The tenant may be given an initial secure period before any break notice can take effect. It is thought that, in many cases, the tenant will be able to obtain a tenant's break option in return for the landlord being given a break option.

Another important exception to the rule that the new tenancy is to be based on the existing tenancy relates to the costs of the new tenancy. Even if the existing lease contains a covenant by the tenant to pay the landlord's costs of granting that lease, each party should normally bear its own costs for the granting of the new tenancy, and such a covenant should not be carried over into the new lease (*Cairnplace Ltd* v *CBL (Property Investment) Co Ltd* [2984] 1 EGLR 59, CA; *Stevenson & Rush Holdings* v *Langdon* [1979] 1 EGLR 72, CA).

Casenotes

O'May v *City of London Real Property Co Ltd* [1982] 1 EGLR 76, HL: The existing tenancy contained a service charge limited to heating and other running costs, the landlord being responsible for the costs of structural repairs, plant renewals and other services. The landlord proposed that the new tenancy should be a "clear lease" imposing a full service charge under which the landlord was fully reimbursed for all his expenditure on the building, this being normal at that time in the open market. The parties' surveyors had agreed that this change, if made, would reduce the new rent by 50p per square foot per annum. The tenant objected. Held: There should be no change on this term of the tenancy. The burden of justifying new terms rested with the landlord who proposed them, and he had failed to show that it was fair and reasonable; the suggested reduction in rent was not adequate to compensate the tenant for taking, under a new lease for only three years, the burden of an unpredictable liability under a fully variable comprehensive service charge.

Gold v *Brighton Corporation* [1956] 3 All ER 442: The tenant used the premises as a shop for the sale of new and second-hand furs and clothes. The council, as landlord, sought to insert in the new tenancy a new clause prohibiting second-hand clothes. Held: The landlord's proposal was rejected since it would adversely affect the tenant's business.

Cardshops Ltd v *Davies* [1971] 2 All ER 721: The existing tenancy permitted assignment subject to landlord's consent which could not be unreasonably withheld. The landlord tried first to impose in the new tenancy an absolute bar on assignment or a proviso for surrender in the event of intended assignment. Held: The landlord's proposals were rejected.

Charles Clements (London) Ltd v *Rank City Wall Ltd* [1978] 1 EGLR 47: The old tenancy limited the use of the premises to a retail cutlers. The landlord sought to add a provision that he would not unreasonably withhold consent for change of use; this would result in a materially higher rent than if the narrow use restriction was repeated. The tenants were content with the old, restrictive use. Held: The landlord's proposal was rejected.

Aldwych Club Ltd v *Copthall Property Ltd* (1962) 185 EG 219: The old lease prohibited use otherwise than as a club without the landlord's consent, which was not to be unreasonably withheld. Before the renewal, the landlord obtained planning permission for change of use to offices. The tenant, who used the premises only as a club, sought to limit the use under the new tenancy by removing the provision for consent not to be unreasonably withheld, thus limiting use to a club only; this would result in a materially lower rent. Held: The tenant's proposal was rejected.

Adams v *Green* [1978] 2 EGLR 46: The landlord owned a parade of 12 shops which were about a hundred years old. At the renewal of a lease of one shop, the landlord proposed a 14 year lease with the right to terminate it for demolition or reconstruction by giving two years' notice at any time. The tenant objected to the break provision. Held: The terms should be as proposed by the landlord.

Amika Motors Ltd v *Colebrook (Holdings) Ltd* [1981] 2 EGLR 62, CA: The landlord of a car showroom obtained planning permission for substantial alterations which would fall within section 30(1)(f). Since his section 25 notice had indicated no opposition to a new tenancy, he could not object to a renewal but proposed during the negotiations with the tenant that the new tenancy should contain a redevelopment break clause on six months' notice. The tenant had recently invested substantially in his car dealership business, was concerned about losing his franchise for the area, and needed sufficient time to find an alternative showroom in the same area; this would be difficult in view of town planning restrictions on car showrooms. Held: The new tenancy should be for five years with a landlord's break clause entitling him to terminate for redevelopment on six months' notice but expiring not earlier than the end of the third year of the new term.

JH Edwards & Sons Ltd v *Central London Commercial Estates Ltd* [1984] 2 EGLR 103: The existing leases of two shops each contained a redevelopment break provision on six months' notice, which had not been exercised. The landlord initially opposed renewals on ground (f), but then entered into an agreement to grant a hotel company a 60 year lease of the block, including the two shops. The hotel company initially proposed to refurbish and occupy the remainder of the buildings, and did not intend to incorporate the shops for some time, so opposition to renewal of the two shop leases was withdrawn. The landlord proposed short or terminable leases but the county court judge ordered 10 and 12 year leases with no redevelopment break clauses. The landlord appealed. Held: The judge's decision unreasonably prevented redevelopment. The new tenancies were to be for seven years determinable for redevelopment as from the end of the fifth year.

National Car Parks Ltd v *Paternoster Consortium Ltd* [1990] 1 EGLR 99: The landlord was proposing to redevelop Paternoster Square and the local planning authority was in favour, but a scheme had not yet been finalised. The landlord originally opposed a renewal on ground (f) but withdrew that opposition and proposed a new 14 year lease with a redevelopment break clause exercisable on six months' notice given at any time. Held: The landlord's proposals were accepted.

12.5 Special terms where the landlord is a public body

In cases where the "public interest" provisions of sections 57 and 58 apply (see 6.9), the court is obliged to order the insertion of a provision that the new tenancy can be terminated by notice to quit given by the landlord, if a certificate is issued under section 57(1) or section 58(1).

In cases where section 60 or 60A applies (see 6.9) the court is obliged to order that any new tenancy shall contain terms prohibiting or restricting the tenant from assigning, subletting, changing or parting with possession of the whole or part of the premises or changing its use, as may be specified in any certificate from the Secretary of State certifying that this is necessary or expedient: section 60(2), section 60A(2).

12.6 Rent under the new tenancy — basic principles

The rent under the new tenancy will be determined by the court if not agreed. The statutory basis of rent valuation as set out in section 34(1) is:

> the rent at which, having regard to the terms of the tenancy (other than those relating to rent) the holding might reasonably be expected to be let in the open market by a willing lessor there being disregarded:
>
> (a) any effect on rent of the fact that the tenant has or his predecessors in title have been in occupation of the holding,
>
> (b) any goodwill attached to the holding by reason of the carrying on thereat of the business of the tenant (whether by him or by a predecessor of his in that business),
>
> (c) any effect on rent of an improvement to which this paragraph applies,
>
> (d) in the case of a holding comprising licensed premises, any addition to its value attributable to the licence, if it appears to the court that having regard to the terms of the current tenancy and any other relevant circumstances the benefit of the licence belongs to the tenant.

The rent is to reflect the impact of the Landlord and Tenant (Covenants) Act 1995, mentioned earlier (section 34(4)).

While it is for the court to assess the rent under the section 34 formula, it must do so by hearing expert evidence from valuers, supported by factual evidence of rents payable under comparable lettings (*Rombus Materials Ltd* v *WT Lamb Properties Ltd* [1999] PLSCS 40, CA). Information which is generally available in the market and which might influence a rental offer is also admissible in evidence (*French* v *Commercial Union Assurance Co plc* [1993] 1 EGLR 113, CA, relating to evidence of the imminent closure of a nearby Sainsbury's supermarket). Rents agreed at rent reviews of comparable properties might be used in support of the surveyors' expert opinion of rental value, but in the knowledge that they may have been affected by the desire of the parties to settle. Rents determined by rent review arbitrations will not normally be admissible in evidence (*Land Securities plc* v *Westminster City Council* [1992] 2 EGLR 15).

There is no direction in section 34 (unlike some of the provisions of section 24A as to interim rent) to have regard to the existing rent paid by the tenant or any other matter under the existing tenancy. The fact that the market position of the area may have become different from that prevailing when the tenant's existing lease was granted is no reason for fixing the new rent at a level different from current open market rent (*Giannoukakis* v *Saltfleet Ltd* [1988] 1 EGLR 73, CA).

If determined by the court, the rent is to be fixed having regard to rent levels current at the date of the court hearing (not the date of termination specified in the section 25 or section 26 notice) with an allowance for any anticipated movement during the following three months and two weeks to the date when the new rent will begin under section 64, discussed in Chapter 10 (*Lovely & Orchard Services Ltd* v *Daejan Investments (Grove Hall) Ltd* [1978] 1 EGLR 44).

Many rent review clauses include phrases or entire provisions copied from section 34 and in some cases the ways in which the courts have interpreted rent review clauses may assist in applying elements of section 34. However this must be approached with caution for two reasons. First, the courts frequently emphasise that each interpretation of a rent review clause will turn on the context of the whole of the individual lease and will not set a precedent for interpreting the same words when found elsewhere. Second, implying an unexpressed provision into a rent review clause in a lease might be justified where it would not be appropriate to imply it into section 34 at a renewal under the Act (*J Murphy & Sons Ltd* v *Railtrack plc* [2002] 2 EGLR 48, CA).

Unlike many rent review clauses, section 34 does not expressly stipulate the hypothesis that the letting would be "to a willing lessee", but in *Murphy* it was accepted that the counterpart in the open market of the willing lessor is a willing lessee (both the willing lessor and the open market being specifically mentioned in section 34(1)). Furthermore, the tenant who is seeking the lease renewal is plainly a willing lessee. Nevertheless, it was decided in *Murphy* that it was not appropriate to imply into section 34, as a consequence of that, the additional implied disregard which had been inferred by the court in two rent review cases, that where the premises are landlocked and can only be accessed through the tenant's other land, one must assess the rent disregarding the landlock aspect. The fact that, at a rent review, the parties may or may not have intended a full undiscounted rent to be determined in that situation is irrelevant to the application of section 34, which is a statutory formula unconnected with the intention of any particular parties.

Another common rent review provision which is absent from section 34 is an assumption that the premises can be let with vacant possession. For the purpose of section 34, it is to be assumed that the tenant has vacated the holding, but where the new tenancy is to comprise the whole premises demised by the current tenancy and parts other than the holding have been sublet, the existence of the subtenancies must not be ignored. If part of the premises is sublet, and the tenant is to be granted a new tenancy which includes the sublet part, the amount of rent payable by him under the new tenancy must take account of the rent he will be able to obtain from the subtenants (*Oscroft* v *Benabo* [1967] 1 QB 1087).

Although section 34 does not expressly state that the premises are assumed to be in the condition that would exist if the tenant had complied with his covenants in his current tenancy, this will be implied since the common law precludes a person from profiting from his own breach of obligation (*Re 5 Panton Street, Haymarket* (1960) 175 EG 49, ChD).

Since section 34 requires the rent to have regard to the terms to be contained in the new tenancy, the rent is to be assessed after the duration and other terms of the new tenancy have been agreed or determined (*Cardshops Ltd* v *Davies* [1971] 1 All ER 721). This enables the rent to reflect, for example, any particularly restrictive user clauses that will be in the new lease. Some examples of the impact on rent of user provisions can be seen from the cases cited below. The actual percentage discounts given in the cases cannot, however, be taken as generally applicable, since each case turns on the valuation evidence before the court. Adjusting for the particular permitted use does not require the court to exclude evidence of comparable transactions where otherwise comparable premises were let on different terms as to use (Hoffmann LJ in the rent review case of *Zubaida* v *Hargreaves* [1995] 1 EGLR 127, CA).

Other aspects of the assessment of the new rent under section 34 are discussed below.

Casenotes

Oscroft v *Benabo* [1967] 1 QB 1087: The tenancy comprised a shop with flat above. The flat was sublet on a controlled tenancy. The new lease was to comprise the entire premises. Held: The new rent for the entire premises had to reflect the controlled rent receivable from the subtenant.

Family Management v *Gray* [1980] 1 EGLR 46: On the expiry of the immediate lease, the subtenants (who had held full repairing subleases) were protected by the 1954 Act and negotiated new full repairing leases directly from the landlord. The landlord then sued his former immediate tenant for dilapidations. Held: The rents obtained from the subtenants under the 1954 Act were assumed to be the full market rent and the lessees could not, in diminution of what was the proper rent, use their own default in having failed to comply with their repairing covenants as a justification for a lower rent, whether arrived at by negotiation or determined by the court. Thus the landlord had suffered no loss.

Plinth Property Investments Ltd v *Mott, Hay & Anderson* [1979] 1 EGLR 17: In this leading case on rent review, the lease contained a covenant that the premises could be used only as offices in connection with the lessee's business of consulting engineers. The arbitrator determining the rent review fixed a rent over 30% below market rent for unrestricted office use. Held: The arbitrator was correct to adjust the rent to reflect the narrow permitted use.

UDS Tailoring Ltd v *BL Holdings Ltd* [1982] 1 EGLR 61: The existing tenancy restricted the use of the premises to tailors and outfitters, and gave the landlord a veto over any change which the landlord considered to conflict with another existing or proposed use in the block. This would be repeated in the new tenancy. Held: The narrow use restriction justified a 10% reduction in rent below unrestricted retail level.

Giannoukakis v *Saltfleet Ltd* [1988] 1 EGLR 73, CA: The tenant had run a snack-bar at the premises for many years. The area became more up-market. Pizzaland opened next door. This was a comparable for assessing the rent and produced a market rent higher than the tenant's old-fashioned business could afford. Held: There is no means under the Act under which this problem can be alleviated.

Re 5 Panton Street, Haymarket (1960) 175 EG 49, ChD: In breach of his repairing covenant, the tenant allowed the basement of the holding to fall into a very bad state of repair. Held: It would not be right to reduce the value of the tenancy of the basement by reason of the poor state of repair.

12.7 Insertion of rent review provisions

The new rent can be subject to review and the court has a discretion as to the review provisions to be inserted into the new tenancy (section 34(3)). The court may have regard to the provisions of the old lease when determining the terms of the review clause, e.g. when deciding whether the review should be "upwards only" or "up/down", although this may not be decisive (*Charles Follett Ltd* v *Cabtell Investments Ltd* [1986] 2 EGLR 76, at first instance). Particularly where the old lease does not contain a rent review clause, the court may order the rent review clause in the new lease to be on an "up/down" basis or "upwards only", or may even omit a review altogether (*Northern Electric plc* v *Addison* [1997] 2 EGLR 111, CA). This issue should be decided before the amount of the initial rent which should reflect the existence and type of review clause. At the time of writing, the Government is considering whether to introduce legislation to prohibit the insertion of upward only provisions into rent review clauses in new leases; if enacted, that would almost certainly affect lease renewals.

There do not appear, at the date of publication, to be any recent reported cases of lease renewals where the existing lease reserved a rent geared to the tenant's turnover, rather than an ordinary market rent review clause. The question arises whether the court would simply import the same turnover rent formula from the existing lease into the new tenancy. Such an approach would be consistent with the *O'May* principle applied to determining the other terms of the new tenancy under section 35 (see above). However, section 35 specifically requires regard to be had to the terms of the old tenancy whereas section 34, which governs determination of the new rent, makes no mention of the terms of the existing lease and at first glance appears to require a fixed initial rent to be determined. Nevertheless in the past the court has sometimes ordered the rent to be determined by a formula on a commission or turnover basis. In *Naylor* v *Uttoxeter District Council* (1974) 231 EG 619, the new rent for a cattle market let to auctioneers was to be a specified percentage of the auctioneer's gross commission. Thus, if section 34 does not in fact require the determination of a fixed amount of initial rent, a full turnover rent based on the provisions of the existing lease might be ordered. Alternatively there might be a fixed basic rent and a turnover-based additional rent determined at frequent intervals.

In *Berthon Boat Co Ltd* v *Hood Sailmakers Ltd* [2000] 1 EGLR 39, the landlord and the tenant had entered into a separate agreement for commission based on sales being paid to the landlord during the lease or any substitute lease, and this arrangement was held on that wording to continue during any continuation under the 1954 Act and any new tenancy.

Casenotes

Janes (Gowns) Ltd v *Harlow Development Corporation* [1980] 1 EGLR 52: The existing tenancy was for 21 years without rent review. The premises were in a parade of shops whose trading was endangered by a forthcoming nearby shopping development. The landlord proposed an upwards-only rent review, but the tenant sought an up/down clause. Held: Because of the danger to trade, and because "what is sauce for the goose is sauce for the gander" (*per* Cross J in *Stylo Shoes Ltd* v *Manchester Royal Exchange Ltd* (1967) 204 EG 803), the review should be up/down.

Blythewood Plant Hire Ltd v *Spiers Ltd* [1992] 2 EGLR 103, Wood Green County Court, Judge Anthony Diamond QC: The tenant had been granted a five year lease of a backland site in 1941 and had remained in possession. The landlord proposed a five year lease with a rent review on the penultimate day. The tenant sought a 14 year lease with five yearly rent reviews on an up/down basis. The parties' surveyors, in evidence, said they had no actual experience of up/down review clauses and until recently these were very rare; it would make the landlord's interest difficult to market; it would be of little immediate benefit to the tenant — a tenant would not pay substantially more rent for an upward/downward clause; and (unlike the *Janes Gowns* case) there was no evidence of nearby development likely to cause the rent to go down. Held: The new lease should be for 10 years with a redevelopment break clause exercisable from the end of the seventh year. There should be five-yearly rent reviews on an upwards-only basis.

Boots the Chemists Ltd v *Pinkland Ltd* [1992] 2 EGLR 98, Wood Green County Court, Judge Thompson: The existing leases of shops did not contain rent review provisions. The landlord sought upwards-only review provisions. The tenant proposed an up/down review basis. Held: The review should be up/down, since this is fair and reasonable, future movements in rent being incapable of prediction. The landlord will not be prejudiced if rent levels increase.

12.8 Disregard of tenant's occupation and goodwill

The direction to disregard tenant's occupation in section 34(1)(a) is a direction to disregard the effect on rent of that occupation. The direction in section 34(1)(b) to disregard any goodwill relates to goodwill attached to the holding by reason of the carrying on there of the tenant's business.

Where the tenant is a company in a group (see 2.7), the direction will encompass occupation by, or goodwill arising from occupation by, the tenant company or by another company in its group (section 42(2)(a)). Where the tenant is a trustee and a beneficiary of the trust is in occupation (see 2.10) the the direction will encompass occupation by, or goodwill arising from occupation by, the beneficiary (section 41(1)(a)).

In relation to cases governed by the 2004 amendments (see 1.2), where the tenant has a controlling interest in a company (see 2.8), these directions will encompass occupation by, or goodwill arising from occupation by, the tenant or by that company, and where the tenant is a company, these directions will encompass occupation by, or goodwill arising from occupation by, the tenant company or by a person with a controlling interest in it (section 23(1B)).

The question arises whether section 34(1)(a) and (b) have the effect of preventing the court from taking account of the success or failure of the actual tenant's trading at the premises. In *Harewood Hotels Ltd* v *Harris* [1958] 1 WLR 108, the Court of Appeal admitted evidence of the tenant's trading at the premises, which were used as an hotel, to show what sort of profit such an hotel might make and what rent a prospective tenant might pay; but in *WJ Barton Ltd* v *Long Acre Securities Ltd* [1982] 1 EGLR 89 the Court of Appeal held that the *Harewood* ruling only applied in limited circumstances and that in general the actual tenant's trading would only indicate whether his business had been successful or not, and had no direct bearing on the amount of the open market rent. A similar issue arose in the rent review cases of *Cornwall Coast Country Club* v *Cardgrange Ltd* [1987] 1 EGLR 146 and *Electricity Supply Nominees Ltd* v *London Clubs Ltd* [1988] 2 EGLR 152, where it was held that information about the actual tenant's trading was not admissible in evidence at a rent review arbitration where the open market rent was to be determined, because that information would not be available to prospective tenants in the open market and so would not affect their rental bids. Under section 34, the rent is similarly to be fixed on the basis of a hypothetical open market letting, and it is thought that the same ruling may apply.

12.9 Disregard of tenant's improvements

The direction to disregard tenant's improvements in section 34(1)(c) and section 34(2) is a direction to disregard the effect on rent of the improvements if they were carried out by a person who at the time was the tenant and were not carried out pursuant to an obligation to his immediate landlord. Further, if they were carried out before the current tenancy began, they must have been carried out not more than 21 years before the court application and, since they were carried out, Part II of the Act must have applied continuously to the holding (or the improved part), and on any expiry of a tenancy during that period there must have been a renewal by the then tenant.

Under the first element of these strict rules, the ambit of section 34(2) does not encompass any improvements made by a person before he became the tenant (for example, during a period of access that had been given to him as a licensee prior to being granted the existing lease) and so such improvements will not be disregarded; any increase in rental value that they justify will be taken in the landlord's favour (*Euston Centre Properties Ltd* v *H&J Wilson* [1982] 1 EGLR 57).

Works carried out by a builder for and at the cost of the tenant will be treated as carried out by the tenant. If works are carried out by or for some other party under an arrangement with the tenant, they may not be treated as carried out by the tenant within the disregarding direction of section 34(2), unless there is sufficient involvement by the tenant in identifying, supervising and/or financing the works, which is a matter of fact and degree and must have regard to a "commercial common sense" view of the arrangements (*Durley House Ltd* v *Viscount Cadogan* [2000] 1 EGLR 60, ChD, Neuberger J).

The question arises as to the meaning of "improvement". At first glance the context appears to indicate that it means alterations which would, if not disregarded, increase the open market rental value of the premises. However, it may be argued that it should also encompass works which the tenant carried out to improve the premises for his particular use, even if they reduced the rental value for tenants generally. On that basis the rent under section 34 would have to be assessed disregarding any increase in rental value or any decrease in rental value resulting from the works, and not just disregarding any increase. In a leading case under section 19 of the Landlord and Tenant Act 1927 in respect of covenants restricting the making of improvements, it was held that "improvements" in that section meant works that improved the premises from the tenant's viewpoint even if they did not improve the rental value (*FW Woolworth & Co Ltd* v *Lambert* [1936] 2 All ER 1523). It is unclear whether that interpretation should also apply to section 34(2) of the 1954 Act; if it does, the disregard will cover both improvements and so-called disimprovements.

The direction in section 34(2) is not to disregard the improvements themselves but to disregard "the effect on rent" of the improvements. How exactly this disregard is to be assessed can give rise to difficulty. Where evidence is available of comparable lettings of similar but unimproved premises, so that a market rent for the unimproved premises can clearly be established, that is to be used as the basis for determining the rent. Where comparable evidence of unimproved premises is not available, the premises must be valued as they stand and an adjustment made to remove the effect on rent of the improvements. Exactly how this is to be done is debatable. For a discussion on the point see the observations of Forbes J in the rent review cases of *GREA Real Property Investments Ltd* v *Williams* [1979] 1 EGLR 121 and *Estate Projects Ltd* v *Greenwich London Borough* [1979] 2 EGLR 85. In the rent review case of *Iceland Frozen Foods plc* v *Starlight Investments Ltd* [1992] 1 EGLR 126 CA, the Court of Appeal decided that the market rent for large premises, which were physically capable of being divided into seven units and for which the tenant had the right under the lease to divide the premises and to sublet in parts, should not reflect the potential for the premises to be subdivided and sublet in parts, since any actual subdivision by the tenant would be disregarded by virtue of the direction to disregard tenant's voluntary improvements and so the potential for subdivision should also be disregarded. This decision was criticised and not followed in *Lewisham Investment Partnership* v *Morgan* [1997] EGLR 126, in which Neuberger J said that even if actual improvements are disregarded, the potential for making them still exists and benefits the tenant, and the rent that a "willing landlord" would accept would have some regard to the rent he might otherwise achieve by dividing the premises and letting them in units, notwithstanding that the rent must be fixed on an assumed letting as a whole. Since section 34, like many rent review clauses, refers to a "willing landlord", this approach may also apply to rent determination under the Act, if it can be shown that such a landlord would realistically expect a tenant in the market would offer a rent which reflected the potential of the premises for improvement. Exactly how this rental value would be assessed is also debatable, but it might be by reference to the rental value which would be released by the improvements compared to their cost. It may be apparent to tenants in the market that by spending, say, £10,000 on improving the premises, the rental value of the premises could be increased by, say, £5,000 per year. In these circumstances it may be fair to reflect in the annual market rent this latent value by taking the sum of £5,000 per year less the cost of amortising the £10,000 expenditure over the term of the new lease and allowing for financing costs. For example, if the new lease is to be for a term of ten years, the expenditure of £10,000 might be amortised at £1,000 per year plus the cost of finance at, say, 7% pa giving £1,700 per year, in which case the additional element of rent reflecting the potential for improvement might be the difference between that and £5,000 per year, namely £3,300 per year.

Casenotes

Euston Centre Properties Ltd v *H&J Wilson* [1982] 1 EGLR 57: The rent review clause in a lease provided for the rent to be assessed "disregarding any of the matters referred to in section 34(a) (b) and (c) of the Landlord and Tenant Act 1954". Before the grant of the lease, the original tenant had voluntarily carried out improvements at their own expense under the terms of an agreement for lease. Held: At that time, the party making the improvements was not a tenant but only a licensee and therefore the improvements must be taken into account and not disregarded.

Daejan Properties Ltd v *Holmes* [1996] EGCS 185: The landlord had granted the tenant a licence to extend the building on the premises. The licence included a positive covenant by the tenant to carry out the specified works and a statement that the works "shall be deemed to be carried out in pursuance of an obligation to the lessor". Held: This had the effect of deeming those works to be obligatory improvements for rent review purposes. (Note: The decision in this rent review case may be applicable to rent determination under section 34 because the wording about improvements is similar).

Durley House Ltd v *Viscount Cadogan* [2000] 09 EG 183, ChD, Neuberger J: A rent review clause in a lease incorporated by reference the provisions of section 34 of the Act. The tenant had entered into a management agreement with a third party that it would refurbish the property to create a luxury aparthotel, manage the business and pay the tenant a percentage of the gross takings. The refurbishment concept, the execution of the works and the standards of operation had to be approved by the tenant. At the rent review under the tenant's lease, the landlord argued that the refurbishment should not be disregarded since it had not been carried out by the tenant within the requirements of section 34(2). Held: The effect of the arrangements between the tenant and the management company were such that on a commercial, common sense view, the improvements had been carried out by the tenant within section 34(2) and were within the disregard.

Amarjee v *Barrowfen Properties* [1993] 2 EGLR 133, Wood Green County Court: The tenant had occupied a furniture warehouse on a yearly tenancy for six years. Apart from the rent of £35,000 pa, there were no written terms. At the renewal under the Act, the parties agreed to a 14 year term but disagreed over many of the terms to be inserted in the new lease, including whether the rent reviews were to be upwards/downwards or upwards only. Held: An upwards/downwards rent review clause was appropriate having the obvious merit of fairness and the only disadvantage was that it could give rise to disputes, in some instances, at the end of review periods.

Boots the Chemists Ltd v *Pinkland Ltd* [1992] 2 EGLR 98, Wood Green County Court, Judge Thompson: The tenant had held an old fashioned 21 year lease with an upwards-only rent review clause but argued for an upwards/downwards clause in the new lease, rents in the area having fallen. Held: In order to do that which is fair and reasonable, the rent should be subject to review downwards as well as upwards. Fixed rents, or rents which can only be revised upwards, can wreak the same sort of injustice upon tenants as that which has been suffered by landlords in previous decades when leases contained no provision for rent review at all.

12.10 Assumed condition of the premises

Although section 34 does not contain an express direction for the court to assume that the premises are in the condition required by the tenant's covenants in his existing lease, the tenant will not be allowed to plead his own default in order to reduce the rent. The premises must be valued as if they were in the condition in which they would have been had the tenant performed his covenants (*Re 5 Panton Street, Haymarket* (1960) 175 EG 49 Ch; *Family Management* v *Gray* [1980] 1 EGLR 46).

This is particularly important where the tenant, in breach of covenant, has allowed the premises to deteriorate without carrying out repairs or redecoration, or has carried out unauthorised alterations which reduce the rental value of the premises (and which, for that reason, might not be "improvements" to be disregarded under section 34(2) mentioned above).

12.11 Licensed premises

The direction in section 34(1)(d) relating to "licensed premises" is to disregard, in the case of a holding comprising licensed premises, any addition to its value attributable to the licence "if it appears to the court that having regard to the terms of the current tenancy and any other relevant circumstances the benefit of the licence belongs to the tenant".

"Licensed premises" is not confined to premises with licences for the sale of alcoholic drinks but includes licensed betting offices (*Ganton House Investments Ltd* v *Crossman Investments Ltd* [1995] 1 EGLR 239, Colchester and Clackton County Court, Judge Brand).

Where the tenant is a company in a group (see 2.7), the direction in section 34(1)(d) will apply whether the licence is held by the tenant company or by another company in the group (section 42(2)(b)). Where the tenant is a trustee (see 2.10), the direction in section 34(1)(d) will apply whether the licence is held by the tenant or by a beneficiary of the trust (section 41(1)(b)).

In cases to which the 2004 amendments apply, where the tenant has a controlling interest in a company, the direction in section 34(1)(d) will apply whether the licence is held by the tenant or that company; and where the tenant is a company, the direction in section 34(1)(d) will apply whether the licence is held by the tenant company or a person with a controlling interest in it (section 34(2A)).

12.12 SDLT

As explained at 4.10, generally SDLT is payable under the Finance Act 2003 on all leases granted on or after 1 December 2003, except where the NPV of the lease is below the threshold; and where SDLT has been paid in respect of the continuation tenancy, the SDLT payable in respect of the new lease may be reduced where the new lease is granted during a year for which that SDLT has been paid.

The way in which an increase in rent for the continuation period, arising from interim rent, may affect the SDLT in respect of the new lease has been explained at 11.6.

A greater amount of SDLT than might otherwise arise might become payable by the tenant if the term of the new lease is expressed, in the new lease, to run from a date before the date on which the new lease is actually granted. Backdating the term will not actually retrospectively commence the term, since the term cannot start until the deed — that is, the new lease — has been executed (see 2.2 and *Bradshaw* v *Pawley* [1979] 3 All ER 273). Nevertheless it will have the effect of making the tenant legally liable under the lease, by virtue of his covenant for payment of rent, to pay the rent for the backdated period. A possible consequence of this is that, if the amount of that rent is greater than the previous passing rent, the amount of the increase in rent for the period before the date of the new lease may be treated, for SDLT purposes, as if it were a premium paid by the tenant for being granted the new lease. That treatment could result in the SDLT payable in respect of the new lease being greater than that which would be payable if the new term is expressed to start only when the new lease is granted and the interim rent is fixed in a separate document, where the SDLT would be calculated on the basis mentioned in 11.6.

13 Agreements and Court Orders for New Tenancies

13.1 Agreement as to certain terms of renewal

The landlord and tenant do not have to go to court to settle the terms of the new tenancy. Each of sections 33 (duration), 34 (rent) and 35 (other terms) provides that the aspect of the new tenancy which it governs can be agreed between the landlord and the tenant.

However, the issue arises as to the degree of finality required to treat any of these matters as having been "agreed". Since section 33 starts with the words "where on an application ... the court makes an order ..." and sections 34 and 35 make similar references to the tenancy being granted by order of the court, it has been said that the particular terms must have been agreed for the purpose of an application which was before the court (Oliver J in *Derby & Co Ltd* v *ITC Pension Trust Ltd* [1978] 1 EGLR 38). That means that the court application must have been made by then and the agreement must have been reached for the purpose of settling that aspect of the proceedings.

If the renewal is the subject of a court hearing and counsel for one party tells the court that a term of the renewal has been agreed, the court might refuse to allow that party to resile from that statement if the hearing proceeds on the basis that the term in question was so agreed (*Boots the Chemists Ltd* v *Pinkland Ltd* [1992] 2 EGLR 98, Judge Thompson, Wood Green County Court).

It is therefore likely that any attempt by the parties to agree any of the terms of renewal prior to the court application will not be effective unless they agree all the terms and enter into an agreement for a new tenancy within section 28, discussed at 13.2.

13.2 Agreement for grant of new tenancy

If an agreement is entered into as to all the terms of a future tenancy of the holding (with or without additional premises), section 28 provides that the Act will cease to apply and the current tenancy will continue until the date on which the new tenancy is to commence. Under section 69(2), all agreements made between the landlord and the tenant for the purposes of the Act must be made in writing. For the purpose of section 28, "agreement" means a binding contractual arrangement enforceable at law (*RJ Stratton Ltd* v *Wallis Tomlin & Co Ltd* [1986] 1 EGLR 104). An agreement reached on a "subject to contract" basis will not be legally binding, even if the parties reach the stage of engrossing an agreed text of the new lease for them to execute. The agreement, to be legally binding, must be free of any such qualification.

Since all the terms for the new tenancy must be agreed so as to bring section 28 into effect, it follows that the agreed terms must include a commencement date for the new tenancy and this must be a date in the "future", a date after the date on which the agreement is entered into. Even if the agreement is that the new term will be expressed to start (and possibly the new rent will be agreed to be payable) from a past date, that agreement should fall within section 28 since the new lease will actually be completed on a future date, and that date will be the date for the new tenancy to commence. This is because a new term cannot be created retrospectively — the provision for a specified length of term to run from an earlier date will only operate to fix the expiry date of the term, and a provision for rent to be payable from an earlier date will create merely a contractual obligation to pay a sum calculated on that basis (*Bradshaw* v *Pawley* [1979] 3 All ER 273). For this reason, any agreement as to the terms of a new tenancy should specify a date for completing the new lease.

For an agreement for a new lease to be legally binding and enforceable, the question arises whether sections 28 and 69 must be read in conjunction with section 2 of the Law of Property (Miscellaneous Provisions) Act 1989. That legislation contains requirements for "any contract for the ... disposition of an interest in land". A "disposition" is defined by reference to the Law of Property Act 1925, in which section 205(1)(ii) states that "disposition" includes conveyance and that "conveyance" includes a lease. "Interest in land" is defined in s.2(6) of the 1989 Act to include any estate or interest in land, and section 1 of the Law of Property Act 1925 states that legal estates in land includes leases for terms certain. Generally, no agreement for the grant of a lease, other than one for a term not exceeding three years at best market rent (which is specifically excluded by section 2(5)(a) of the 1989 Act), can be validly made without complying with the formal requirements of section 2. These are that the " contract ... can only be made in writing and only by incorporating all the terms which the parties have expressly agreed in one document or, where contracts are exchanged, in each" (section 2(1)); and "the document incorporating the terms or, where contracts are exchanged, one of the documents incorporating them (but not necessarily the same one) must be signed by or on behalf of each party to the contract" (section 2(3)). This is supplemented by a provision in section 2(2) that "the terms may be incorporated in a document either by being set out in it or by reference to some other document".

Where the landlord's title is registered at HM Land Registry, an agreement held by a tenant in occupation, relating to the premises he occupies, is likely to be protected automatically as an interest which overrides a transfer by falling within schedule 3 para 2 of the Land Registration Act 2003 as an "interest belonging ... to a person in actual occupation ... relating to land of which he is in actual occupation...". Such overriding interests prevail against a transferee of the landlord's title by section 29 of the 2003 Act. However, if the agreement is for a new tenancy which will include premises other than those currently occupied by the tenant, it may not be protected in relation to those other premises unless it is entered on the register of the landlord's title to the premises affected.

Where the landlord's title is unregistered, the agreement should be registered at the Land Charges Department as an estate contract, class C(iv), against the name of the estate owner, ie the landlord. Failure to do so might render the agreement void against a purchaser of the landlord's interest by virtue of section 198 of the Law of Property Act 1925 and section 4(6) of the Land Charges Act 1972.

It is generally the rule that a legally binding contract for a lease cannot be created by mere exchange of correspondence. Unless the view is taken that section 28 and section 69(2) override the general law, the safest procedure (if there is some reason for wishing to conclude a binding agreement rather than proceeding directly to completion of the new lease) is either to exchange a formal agreement for lease or to have the agreed terms embodied in a consent order of the court, in which the court orders a new tenancy to be granted on those agreed terms, reserving to either party liberty to apply to the court to determine any part of the text of the new lease that cannot subsequently be agreed. This should prevent one party or the other reneging on the basic terms of the deal.

In the event of a delay between concluding the agreement for a new lease and completing the grant of the new lease, the tenant should consider protecting the agreement by title registration, as mentioned above.

Casenotes

Derby & Co Ltd v *ITC Pension Trust Ltd* [1978] 1 EGLR 38: The tenant's agents wrote to the landlords confirming the agreed terms for a new tenancy but their letter was marked "subject to contract" and they expressly stated that the agreement was without prejudice to the tenant's rights under the 1954 Act procedures. The tenant refused to execute the engrossment of the agreed form of new lease and sought to renegotiate the terms. Held: There had been no legally binding agreement.

Salomon v *Akiens* [1993] 1 EGLR 101, CA: After a section 25 notice was served there was correspondence between the parties, agreeing the terms for a new lease. It was marked "subject to contract". The tenant made no application to the court. The form of lease was agreed but the landlord refused to grant it, since arrears of rent were unpaid, and sought possession. Held: The landlord was entitled to possession. There was no binding agreement for lease. The landlord was not estopped from denying the absence of a binding agreement.

RJ Stratton Ltd v *Wallis Tomlin & Co Ltd* [1988] 1 EGLR 104: The tenants agreed terms for a new tenancy with their landlord. The landlord then sold his reversion before the new lease was granted. The tenants had to argue that the agreement was not legally binding, since otherwise section 28 would deprive them of their right to apply to the court for a new tenancy, and their failure to register the agreement as a land charge (their landlord's interest did not have a registered title) would mean that it did not bind their new landlord. Held: It had been a binding agreement and was caught by section 28. The court did not have jurisdiction to hear their application for a new tenancy.

13.3 Implementation of court orders

Where the court makes an order under section 29 for the grant of a new tenancy (either by consent on terms agreed by the parties or on terms determined by the court under sections 33, 34 and/or 35), then section 36(1) provides that the landlord is obliged to execute and the tenant is bound to accept (and, if the landlord requires, execute a counterpart of) a lease or tenancy agreement embodying those terms, unless the order is revoked at the tenant's request under section 36(2) (see 13.4).

13.4 Tenant's right to have court order revoked

If the tenant does not want to take up the new tenancy ordered by the court, he can apply to the court under section 36(2) to have the order revoked. To do so, the tenant must apply to the court to do so within 14 days after the making of the order for the grant of the new tenancy.

If an order for a new tenancy is revoked under section 36(2), any award of costs included in the order will survive the revocation, but the court has a discretion to revoke or vary any such provision as to costs or, where no award was previously made, to make an award of costs against the tenant for the landlord's abortive expenditure.

In the event of the court order being revoked, the current tenancy continues under section 64 (see 10.23) automatically for three months from the date of the order. This continuation will be on the terms discussed in Chapter 4, and will be at the rent last payable under the existing lease or at an interim rent if one is agreed or determined (see Chapter 11). Furthermore the court has power, under section 36(2), to order that the tenancy will continue beyond that date for such period as may be agreed by the landlord and the tenant, or as may be determined by the court to be necessary to afford the landlord a reasonable opportunity for reletting or otherwise disposing of the premises. That continuation of the tenancy will be on the same terms as the automatic continuation (including the rent, which might be the old contracted rent or an interim rent), but will not carry further protection under the Act.

Compensation for Non-Renewal

14

14.1 Entitlement to statutory compensation

Subject to the limited right of the landlord to exclude compensation by a provision of the tenancy, discussed below, a tenant whose tenancy is protected by the Act will be entitled to compensation under section 37 when he vacates the property, if the court refuses to order the grant of a new tenancy because of opposition by the landlord on ground (e), (f) or (g) of section 30(1) of the Act (explained in Chapter 8) and on no other grounds.

The compensation can also be claimed where the matter was not decided by the court, if the tenant vacates the property having refrained for making a court application for a new tenancy, or having withdrawn his court application, or (in cases to which the 2004 amendments apply) having consented to the landlord's court application being withdrawn, and in each case the landlord had specified one or more of grounds (e), (f) or (g) for opposing a new tenancy, but not any of grounds (a) to (d), in the section 25 notice or in the counter-notice to a section 26 request. This right to compensation generally applies even if the landlord, having served such a notice of opposition to renewal, changes his mind and states that he is withdrawing that opposition (see 8.10) — the tenant may still vacate and claim compensation. In a recent case, the landlord who had changed his mind about redeveloping the property tried to get around this rule by applying to the court under section 29(2) for an order for a termination of the current tenancy while conceding that he could not establish the necessary ground to succeed on that application. His intended tactic was that the court would have to order the grant of a new tenancy, sidestepping the tenant's right to compensation under section 37 which does not apply if a new tenancy is ordered by the court. The tenant successfully applied to have the landlord's application struck out as an abuse of the process (*Felber Jucker & Co Ltd* v *Sabreleague Ltd* [2005] PLSCS 162, ChD, Master Moncaster). This case was decided merely by a Master in the High Court, but it is hoped that a higher court would uphold that decision which is plainly consistent with the intention of the legislation — the landlord's strategy was possible only due to the 2004 amendments allowing court applications to be made by landlords as well as tenants and making what were intended, in this context, only to be consequential amendments to section 37.

The old version of section 37 applies to cases governed by the Act prior to the 2004 amendments (see Chapter 1) but it is in substance very similar to the new version, except for the differences mentioned later.

The new version of section 37 refers to grounds (e), (f) and (g) as the "compensation grounds" and frames the right to compensation as arising in three "compensation cases". The first compensation case is where the court refuses a tenant's application for a new tenancy, made under section 24(1), because of one or more of the compensation grounds but no other ground (section 37(1a)). The second compensation case is where the landlord successfully makes, under section 29(2), an application for an order for termination of the current tenancy without the grant of a new tenancy, on one or more of the compensation grounds but not any other ground (section 37(1B)). In either of those cases, the tenant can ask the court to certify that its decision was based on the satisfaction of the relevant compensation ground (section 37(4)). The third compensation case is where the landlord states, in a section 25 notice or in a counter-notice to a section 26 request, that he is opposed to the grant of a new tenancy on one or more of the compensation grounds but no other ground, and either no application to the court is made under section 24(1) or section 29(2) or such an application is made but is subsequently withdrawn (section 37(1C)).

If the landlord's grounds of opposition include one or more of grounds (a) to (d) of section 30(1) (see Chapter 8), no compensation will be payable unless the court expressly rejects that ground or the landlord withdraws it. This is because grounds (a) to (d) involve defaults by the tenant, and if one of those grounds is proved by the landlord, or accepted by the tenant, it would be unwarranted to require the landlord to pay compensation to the tenant for not renewing his tenancy. If a tenant is faced with the landlord opposing renewal on one or more of grounds (a) to (d) as well as on one or more of the compensation grounds, the tenant should try to persuade the landlord to withdraw that element of his opposition which is based on grounds (a) to (d), failing which the tenant should in appropriate cases challenge any such ground in court, whether or not he decides to challenge the landlord on his compensation ground. If the tenant, by default, concedes one or more of grounds (a) to (d), he cannot claim compensation even if the landlord has also opposed renewal on a compensation ground.

The tenant will become entitled to compensation only on quitting the "holding" (see Chapter 5). The right to compensation crystallises at the date of quitting the holding and that is generally the date on which the position is to be assessed (*Cardshops Ltd* v *John Lewis Properties Ltd* [1982] 2 EGLR 53, CA, approving *International Military Services Ltd* v *Capital & Counties plc* [1982] 1 EGLR 71). While the concept of a "holding" under section 23(3) appears to be limited to the situation where there is a tenancy protected by the Act, the view has been expressed, but not determinatively, that compensation can be claimed by a person who had such a tenancy even if it ended before the time when he quits the premises and he has been continuing in occupation in the intervening period not as a tenant but as merely an unprotected tenant at will (Pumfrey J in *London Baggage Co (Charing Cross) Ltd* v *Railtrack plc (No 2)* [2003] 1 EGLR 141, ChD).

The fact that the tenant may not want a new tenancy does not deprive him of the right to compensation (*Lloyds Bank Ltd* v *National Westminster Bank Ltd* [1982] 1 EGLR 83, cited in *Sun Life Assurance plc* v *Thales Tracs Ltd* [2001] 2 EGLR 57, CA).

In cases where the partnership rules apply (see 2.9), the "business tenants" are the persons entitled to claim the compensation (section 41A(7)).

14.2 Exclusion of entitlement to compensation

Compensation will be wholly excluded if the tenancy itself has been fully contracted out of the protection of the Act (see 3.2). This is because contracting out involves excluding the operation of sections 24 to 28, and a claim to compensation under section 37 has to be triggered by a notice under section 25 or a counter-notice under section 26, both of which fall within sections 24 to 28.

Alternatively, the right to claim compensation can be excluded by agreement, often contained as a term of the existing tenancy. However, such a provision will not be effective if, by the date when the tenant quits, the tenant together with any predecessor in the same business has occupied the holding for the purpose of that business, or for that and other purposes, for at least five years (section 38(2)).

In measuring the five year period, any period of occupation by the tenant as a licencee or tenant-at-will is to be ignored since the five year period must be a period during which the holding has been occupied under a tenancy to which the Act applies (*London Baggage Co (Charing Cross) Ltd* v *Railtrack plc (No 2)* [2003] 1 EGLR 141, ChD).

Where the tenancy is within the Act due to the operation of the rules relating to group companies (see 2.7), an assignment of the tenancy from one company to another in the group will be ignored for this purpose and will not interrupt a five year period (section 42(2)(c)). Where the trust rules apply (see 2.10), an assignment of the tenancy from one trustee to another will similarly be ignored (section 41(1)(c)). Furthermore, since the occupation by and business of any group company of the tenant, or any company in which the tenant has a controlling interest, or any beneficiary of the trust for which the tenant holds the tenancy, will in each case count as the occupation by and the business of the tenant (respectively section 42(2), section 23(1B) and section 41(1)(a)), it seems that a change in the occupier from one company or person in the relevant category under those provisions to another person in the same category will equally be treated as not disturbing the continuity of occupation and business for the five year rule. Where the tenancy is held for Government purposes (see 2.21), on any change of Government occupier there is a deemed succession to the business if the test of occupation for the purposes of a Government department is met both before and after the change of occupier, and such a change will be treated as not disturbing the continuity of occupation and business for the five year rule (section 56(3)).

14.3 Calculation of statutory compensation

The calculation of the compensation is governed by section 37, which has been amended by schedule 7 of the Local Government and Housing Act 1989 and also by the RRO. The standard basis of assessing the amount of compensation is by multiplying the rateable value of the holding by a multiplier prescribed by regulations. Currently the multiplier is governed by the Landlord and Tenant Act 1954 (Appropriate Multiplier) Order 1990, SI 1990 No 363, under which the standard multiplier is one, so that the standard compensation will be equal to the rateable value of the holding. The rateable value to be used in this calculation is the rateable value of the holding in the Valuation List at the date the section 25 notice was served or the date of the counter-notice by the landlord to a section 26 request (*Plessey & Co Ltd* v *Eagle Pension Funds Ltd* [1990] 2 EGLR 200). However, where all or part of the premises are residential, there is a special rule mentioned below.

Where the "holding" has no separate rateable value or where there are several assessments of different parts, the rateable values are aggregated or apportioned, and in the absence of a rateable value a proper assessment is to be used; any dispute on these issues is to be referred to the Commissioners of Inland Revenue for decision by a valuation officer (section 37(5)).

Prior to 31 March 1990, rateable values were given to both residential and commercial premises. However, as from 1 April 1990 rateable values ceased to be given to premises which for rating purposes were classified as residential hereditaments. This would have presented a problem for calculating the statutory compensation under section 37 in respect of holdings which included residential premises, for example where a tenancy comprises a shop with residential flat above. As a consequence, schedule 7 of the 1989 Act amended section 37 by providing that where part of the holding is "domestic property" as defined in section 66 of the Local Government Finance Act 1988, the compensation is the aggregate of compensation calculated on the normal rules set out above on the parts of the holding other than the domestic property (ie disregarding the domestic property when determining the rateable value on which to base the compensation) and if that domestic property was occupied by the tenant, an amount for the tenant's "reasonable expenses in moving from the domestic property" (section 37(5A)). If, unusually, the whole of the holding is "domestic property" (eg where the tenant's business is the provision of lodgings), the compensation is to be based on a notional rateable value equal to the rent at which the premises might reasonably be expected to be let from year to year if the tenant undertook liability to pay all usual tenant's rates and taxes and to bear the cost of repairs and insurance and any other expenses necessary to maintain the holding in a state to command that rent (section 37(5C)). This value is to be determined as at the date of giving of the landlord's section 25 notice or his section 26(6) counter-notice, and any dispute is to be referred to the Commissioners of Inland Revenue for determination by a valuation officer, with an appeal to the Lands Tribunal (section 37(5D)).

Where the landlord's reversion has been divided between two or more landlords and they collectively (see Chapter 5) oppose a renewal of the tenancy on one or more of the compensation grounds, their liability to pay compensation under section 37 is to be calculated separately for each landlord's part of the holding and recovered only from the appropriate landlord (section 37(3B)). However these provisions do not state who is to make the calculation, so it would ultimately have to be determined by the court if not agreed.

14.4 Double compensation

In some cases the tenant will be entitled to double the standard amount of compensation. This will be where there has been continuous occupation of at least part of the holding by the tenant or his predecessor in the business for a full period of 14 years ending on the date of termination of the tenancy (section 37(3)). This is an area where the law applicable to cases to which the 2004 amendments apply differs materially from the law which applies to earlier cases.

Where the law prior to the 2004 amendments applies (see Chapter 1), occupying any part of the holding for the full 14 years or more entitles the tenant to double the standard amount of compensation for the whole holding (*Edicron Ltd* v *William Whiteley Ltd* [1983] 2 EGLR 81, CA).

For cases where the 2004 amendments apply, if the 14 year test is met for the whole holding, the entire amount of compensation is doubled (section 37)(2)), but if it is met only as to part of the holding, only the compensation for that part is doubled (section 37(3A). Unfortunately, there is no provision giving the valuation officer the function of allocating a rateable value between different parts of the holding; the powers given by section 37(5) mentioned above apply only for the purpose of section 37(2) and not also for the purpose of section 37(3A). Accordingly, it would presumably be a matter for the court to determine the calculation of compensation where double compensation is payable in relation to part only of the holding and that part does not have a separate rateable value.

For the purpose of assessing whether there has been continuous occupation for the purpose of the tenant's business for the full 14 years required to qualify for double compensation, an assignment of the tenancy will not matter if the assignee takes over the assigning tenant's business as a going concern, since it is the continuity of the business which counts. However, if the tenancy is assigned but the assignee does not take over the assignor's business, there will be no continuity for the 14 year test. This will be a question of fact, and an express purchase of the previous tenant's goodwill may not be essential in determining whether the business was continued.

Where the provisions of section 42 as to groups of companies apply (see 2.7), an assignment of the tenancy from one group member to another is not treated as a change of tenant for measuring the period of occupation for compensation purposes (section 42(2)(c)). Where the provisions of section 41 as to trusts apply (see 2.10), a change in the persons who are the trustees holding the tenancy is similarly treated as not involving a change of tenant for this purpose (s.41(1)(c)). Furthermore, since the occupation by and business of any company in which the tenant has a controlling interest will count as the occupation by and the business of the tenant (section 23(1B)), it seems that a change in the occupier from one company controlled by the tenant to another company controlled by the tenant will equally be treated as not disturbing the continuity of occupation and business for the 14 year rule. Where the tenancy is held for Government purposes (see 2.21), on any change of Government occupier there is a deemed succession to the business if the test of occupation for the purposes of a Government department is met both before and after the change of occupier, and such a change will be treated as not disturbing the continuity of occupation and business for the 14 year rule (section 56(3)).

The date of termination of the tenancy for measuring the period of occupation is the date for termination specified in the landlord's section 25 notice or in the tenant's section 26 request for a new tenancy, and one must count backwards from that date. Where the tenant initially occupied the property as a licensee for the purpose of fitting it out for his business, prior to his tenancy commencing, that prior period does not count towards the 14 years since at that time there was no tenant and no holding (*Secretary of State for the Environment* v *Royal Insurance Co Ltd* [1987] 1 EGLR 83).

The requirement for 14 years continuous occupation for the purpose of the vacating tenant's business, down to that termination date, means that if the tenant vacates before the termination date he may lose his right to the doubling of the compensation (*Sight & Sound Education Ltd* v *Books etc Ltd* [1999] 3 EGLR 45, ChD), although vacating for just a short period for the purpose of cleaning the premises for handing back to the landlord on the due expiry date may be treated as part of the period of continued occupancy (*Bacchiocchi* v *Academic Agency Ltd* [1998] 2 All ER 241, CA).

15 Compensation for Improvements

15.1 Entitlement to compensation for improvements

Part I of the Landlord and Tenant Act 1927 contains provisions which may entitle a tenant, when he quits the premises at the end of his tenancy, to compensation from the landlord for "improvements" to the premises. All the following references in 15.1 to sections are to sections of the 1927 Act unless otherwise stated.

Section 17 of the 1927 Act provides that the provisions in Part I of the 1927 Act apply to premises held under a lease if they are used wholly or partly for the carrying on at the premises of any "trade or business". If the premises are used for a "profession" rather than for a "trade or business", they will be treated as being for a trade or business within Part I of the 1927 Act if they are used "regularly" for carrying on that profession, and even then it is only the provisions as to improvements in Part I of the 1927 Act that will apply (section 17(3)). Part I of the 1927 Act also contains other provisions, outside the scope of this book.

Mining leases, leases constituting an agricultural holding under the Agricultural Holdings Act 1986 and holdings held under farm business leases within the Agricultural Tenancies Act 1995 (see 2.18) are excluded from Part I of the 1927 Act (section 17(1)). Also excluded are tenancies granted to a person as the holder of an office, appointment or employment from the landlord but, in the case of tenancies created after the commencement of the 1927 Act, only if they are made in writing and expressly state the purpose for which they were granted (section 17(2)).

Where premises are used partly for a trade or business (within the above meaning) and partly for other purposes, only improvements made to the parts used for trade or business will be covered by Part I of the 1927 Act (section 17(4)).

There can be no claim for compensation if the improvements were made before 25 March 1928 (section 2(1)(a)).

There can be no claim if the improvements were carried out under an obligation pursuant to a contract (including a building lease) entered into for valuable consideration (section 2(1)(b)).

In order to claim compensation, the tenant must quit the holding. If he accepts a new lease then he cannot claim compensation, but the effect on rent of the improvements will be disregarded when the new rent is fixed (see 12.6).

15.2 Meaning of "improvements"

The 1927 Act does not expressly define the word "improvements", but in order that a successful claim may be made for compensation, the improvement must be something that "adds to the letting value of the holding" (section 1(1)). The effect of this is that work which has made the premises more useful to the actual tenant but which does not add to the rental value of the premises in the open market does not rank as an improvement for the purposes of compensation under the 1927 Act. Furthermore, compensation cannot be claimed for certain improvements, mentioned below.

An improvement can be extremely substantial, and even includes the erection of a building or the demolition of an existing building and its replacement by a new building. Work will not be an improvement if it falls within the tenant's obligation to repair, nor does improvement include the provision of a tenant's or trade fixture which the tenant is by law entitled to remove. An improvement must effect a physical change in the holding and therefore (for example) the mere obtaining of planning permission for a more valuable use cannot qualify as an improvement.

If improvements were made in pursuance of a statutory obligation, then compensation is payable only if those improvements were carried out after 1 October 1954 (section 2(1)(b)).

15.3 Pre-conditions for the claim

In order to be entitled to claim compensation for improvements, the tenant must have complied with the procedure laid down by the 1927 Act before he carried out the improvements (section 3(5)). This procedure is set out in section 3(1) to (4). Before carrying out the improvement, the tenant must have served on the landlord notice of his intention to make it, accompanied by a specification and plan showing the proposed improvement. There is no prescribed form of notice and a letter would have sufficed if it was made clear that it was given under section 3(1) (*Deerfield Travel Services Ltd* v *Leathersellers' Company* [1982] 2 EGLR 39, CA).

A notice will not be invalid merely because the description of works included items of repair in addition to improvements. The notice will equally not have been invalidated if the plan and specification were served separately from the notice, and the specification itself can have been printed on the plan. The notice and documentation must have been sufficiently detailed to allow the landlord to make the decision whether or not to exercise his option under the 1927 Act of carrying out the works for a reasonable increase in rent.

If the landlord did not serve on the tenant a notice of objection to the works within three months from the request for consent or (if later) from the date on which the landlord received fully detailed plans, and did not offer to carry them out himself, the works will have been deemed to be authorised by the landlord and the tenant could proceed to execute them but only strictly according to his deposited plan and specification (section 3(4)). If the landlord objected within the three month period, the improvements could not have been lawfully carried out unless the objection was withdrawn or the tenant obtained an order of the court authorising the work. Any such order can only be obtained before the work is carried out (*Hogarth Health Club Ltd* v *Westbourne Improvements* [1990] 1 EGLR 89, CA). Obtaining the order requires the tenant to satisfy the court that the works meet the criteria set out in section 3(1), which are that the improvement

(a) is of such a nature as to be calculated to add to the letting value of the holding at the termination of the tenancy; and
(b) is suitable and reasonable to the character thereof; and

(c) will not diminish the value of any other property belonging to the same landlord or to any superior landlord from whom the immediate landlord of the tenant directly or indirectly holds.

If improvements were made in pursuance of a statutory obligation, the tenant would still have had to serve on his landlord a notice of intention to make the improvements together with a specification and plan, but the landlord will have had no right to object and the tenant need not have obtained a certificate that the improvement was a proper one, but he can if he desires apply to the court for a certificate that the improvement has been duly executed (section 2(1)(b) of the 1927 Act as amended by section 48(1) of the 1954 Act).

15.4 Making the claim for compensation

The claim for compensation must be made in the time and manner prescribed by section 41 of the 1954 Act. Where a tenancy has been terminated by notice to quit (including a section 25 notice under the 1954 Act) the period for claiming is three months beginning on the date on which the notice is given. Where the tenant has served a section 26 request under the 1954 Act, the claim must be made within three months from the date on which the landlord gives a counter-notice or, if he has not given such a notice, the latest date on which he could have given a counter-notice under section 26(6), being two months after the giving of the section 26 request. Where a tenancy comes to an end by effluxion of time, the compensation must be claimed not earlier than six nor later than three months before the termination of the tenancy. If a tenancy is terminated by forfeiture or re-entry, the compensation must be claimed within the period of three months beginning with the effective date of the order of the court for recovery of possession of the land or within three months beginning with the date of re-entry if there was no court order — the effective date of an order is the date when an order is to take effect according to its terms, or the date on which it ceases to be subject to appeal, whichever is the later. If there is a surrender of a tenancy, the right to compensation will be lost and therefore the compensation must be bargained for, as part of the terms of surrender.

The claim for compensation must be in writing, signed by the claimant, his solicitor or agent, stating the name and address of the claimant and of the landlord against whom the claim is made and containing a description of the holding and of the trade or business carried on there. Also it must contain a concise statement of the nature of the claim, particulars of the improvements including the date on which it was completed and its cost, and a statement of the amount of compensation claimed.

The improvements must have been made by the claiming tenant or his predecessor in title, which in this case means predecessor in title to the leasehold interest (section 1(1)). Improvements carried out by a subtenant are not included unless the sub-tenant has himself followed the appropriate procedure, and his claim is against his "landlord" (section 1(1)), who can claim reimbursement from the superior landlord provided that he follows the procedure set down in the Act (section 8(1)).

15.5 Calculating the compensation

The amount of compensation is governed by section 1 of the 1927 Act and is a sum equal to the net addition to the value of the holding as a whole as the direct result of the improvement (ie the value of the reversion if the improvements were carried out less its value if they were not but recognising the potential for the premises to be improved) or, if lower, a sum equal to the reasonable cost of carrying out the improvement at the termination of the tenancy less an amount equal to the cost (if any) of putting the work constituting the improvement into a reasonable state of repair, except to the extent that this cost is covered by the tenant's liability under the repairing covenants.

The value is to be assessed by the court if the parties cannot agree (section 1(4)) and the assessment must have regard to any change of use intended by the landlord after the termination of the tenancy (section 1(3)). The implication is that the compensation is to be based on the value of the improvements to the landlord having regard to the actual circumstances. If the landlord intends to demolish or structurally alter the premises after the termination of the holding, regard must be had to this when assessing the value of the improvements. Thus where the landlord intends to demolish the premises, or carry out structural alterations which would negate the improvements, after the termination of the tenancy, there will be no compensation. If the landlord represents that he will carry out such work and so avoids paying compensation, the tenant can make a further application to the court to reassess the compensation claim if the landlord does not in fact carry out the work (section 1(4)).

Glossary

16

16.1 Holding

This is defined in section 23(3), and means the premises comprised in the tenancy excluding any part which is not actually occupied by the tenant or by a person employed by him for the purpose of the business by reason of which the tenant has the protection of the Act.

In other words, it consists of the parts of the premises comprised in the tenancy which are occupied by the tenant. It includes any parts which are occupied by a person employed by the tenant for the purposes of the tenant's business carried on in any part of the premises. It excludes any parts of the premises which the tenant has sublet or has left unoccupied.

16.2 Landlord

This is defined in section 44(1). In most provisions of the Act, the expression "the landlord" means, in relation to a particular tenant, the person having a freehold or leasehold interest in the premises which is superior to that tenant's tenancy and who is that tenant's "competent landlord" (see 16.3), whether being the immediate landlord or a superior landlord.

However, in the provisions relating to landlords giving notices to quit (see 4.5) it means the tenant's immediate landlord (section 44(2)).

Where different landlords own different parts of the premises let on a single tenancy, the "landlord" means all of them acting collectively (section 44(1A) and see 5.1).

16.3 Competent landlord

In the context of a subtenancy, this means the person who, as the subtenant's immediate or superior landlord, is in a position to deal with him in respect of the procedures under the Act. The rules for identifying the competent landlord are set out in section 44(1) of the Act and discussed at 9.2.

16.4 Mesne landlord

In the context of a subtenancy, this generally means the subtenant's immediate landlord (see 9.2) but in some cases there may be more layers of mesne landlords between the relevant tenant and his competent landlord.

16.5 Track

This is the court's tracking system under which claims fall into three categories: small claims, fast track and multitrack. Fast track applies where the court hearing is likely to last only a day or two. Multitrack is for more complex cases.

Appendix 1

Landlord and Tenant Act 1954 Part II Showing Amendments

PART II

SECURITY OF TENURE FOR BUSINESS, PROFESSIONAL AND OTHER TENANTS

Tenancies to which Part II applies

Tenancies to which Part II applies

23-(1) Subject to the provisions of this Act, this Part of this Act applies to any tenancy where the property comprised in the tenancy is or includes premises which are occupied by the tenant and are so occupied for the purposes of a business carried on by him or for those and other purposes.

(1A) Occupation or the carrying on of a business—

(a) by a company in which the tenant has a controlling interest; or

(b) where the tenant is a company, by a person with a controlling interest in the company,

shall be treated for the purposes of this section as equivalent to occupation or, as the case may be, the carrying on of a business by the tenant.

(1B) Accordingly references (however expressed) in this Part of this Act to the business of, or to use, occupation or enjoyment by, the tenant shall be construed as including references to the business of, or to use, occupation or enjoyment by, a company falling within subsection (1A)(a) above or a person falling within subsection (1A)(b) above.

(2) In this Part of this Act the expression "business" includes a trade, profession or employment and includes any activity carried on by a body of persons, whether corporate or unincorporate.

(3) In the following provisions of this Part of this Act the expression "the holding," in relation to a tenancy to which this Part of this Act applies, means the property comprised in the tenancy, there being excluded any part thereof which is occupied neither by the tenant nor by a person employed by the tenant and so employed for the purposes of a business by reason of which the tenancy is one to which this Part of this Act applies.

(4) Where the tenant is carrying on a business, in all or any part of the property comprised in a tenancy, in breach of a prohibition (however expressed) of use for business purposes which subsists under the terms of the tenancy and extends to the whole of that property, this Part of this Act shall not apply to the tenancy unless the immediate landlord or his predecessor in title has consented to the breach or the immediate landlord has acquiesced therein.

In this subsection the reference to a prohibition of use for business purposes does not include a prohibition of use for the purposes of a specified business, or of use for purposes of any but a specified business, but save as aforesaid includes a prohibition of use for the purposes of some one or more only of the classes of business specified in the definition of that expression in subsection (2) of this section.

Continuation and renewal of tenancies

Continuation of tenancies to which Part II applies and grant of new tenancies

24–(1) A tenancy to which this Part of this Act applies shall not come to an end unless terminated in accordance with the provisions of this Part of this Act; and, subject to the ~~provisions of section twenty-nine of this Act, the tenant under such a tenancy may apply to the court for~~ **following provisions of this Act either the tenant or the landlord under such a tenancy may apply to the court for an order for the grant of** a new tenancy—

(a) if the landlord has given notice under section 25 of this Act to terminate the tenancy, or

(b) if the tenant has made a request for a new tenancy in accordance with section 26 of this Act.

(2) The last foregoing subsection shall not prevent the coming to an end of a tenancy by notice to quit given by the tenant, by surrender or forfeiture, or by the forfeiture of a superior tenancy unless—

(a) in the case of a notice to quit, the notice was given before the tenant had been in occupation in right of the tenancy for one month; or

~~(b) in the case of an instrument of surrender, the instrument was executed before, or was executed in pursuance of an agreement made before, the tenant had been in occupation in right of the tenancy for one month.~~

(2A) Neither the tenant nor the landlord may make an application under subsection (1) above if the other has made such an application and the application has been served.

(2B) Neither the tenant nor the landlord may make such an application if the landlord has made an application under section 29(2) of this Act and the application has been served.

(2C) The landlord may not withdraw an application under subsection (1) above unless the tenant consents to its withdrawal.

(3) Notwithstanding anything in subsection (1) of this section—

(a) where a tenancy to which this Part of this Act applies ceases to be such a tenancy, it shall not come to an end by reason only of the cesser, but if it was granted for a term of years certain and has been continued by subsection (1) of this section then (without prejudice to the termination thereof in accordance with any terms of the tenancy) it may be terminated by not less than three nor more than six months' notice in writing given by the landlord to the tenant;

(b) where, at a time when a tenancy is not one to which this Part of this Act applies, the landlord gives notice to quit, the operation of the notice shall not be affected by reason that the tenancy becomes one to which this Part of this Act applies after the giving of the notice.

~~**24A** (1) The landlord of a tenancy to which this Part of this Act applies may,~~

~~(a) if he has given notice under section 25 of this Act to terminate the tenancy; or~~

~~(b) if the tenant has made a request for a new tenancy in accordance with section 26 of this Act;~~

~~apply to the court to determine a rent which it would be reasonable for the tenant to pay while the tenancy continues by virtue of section 24 of this Act, and the court may determine a rent accordingly.~~

~~(2) A rent determined in proceedings under this section shall be deemed to be the rent payable under the tenancy from the date on which the proceedings were commenced or the date specified in the landlord's notice or the tenant's request, whichever is the later.~~

~~(3) In determining a rent under this section the court shall have regard to the rent payable under the terms of the tenancy, but otherwise subsections (1) and (2) of section 34 of this Act shall apply to the determination as they would apply to the determination of a rent under that section if a new tenancy from year to year of the whole of the property comprised in the tenancy were granted to the tenant by order of the court.~~

Applications for determination of interim rent while tenancy continues

24A.–(1) Subject to subsection (2) below, if—

(a) the landlord of a tenancy to which this Part of this Act applies has given notice under section 25 of this Act to terminate the tenancy; or

(b) the tenant of such a tenancy has made a request for a new tenancy in accordance with section 26 of this Act,

either of them may make an application to the court to determine a rent (an "interim rent") which the tenant is to pay while the tenancy ("the relevant tenancy") continues by virtue of section 24 of this Act and the court may order payment of an interim rent in accordance with section 24C or 24D of this Act.

(2) Neither the tenant nor the landlord may make an application under subsection (1) above if the other has made such an application and has not withdrawn it.

(3) No application shall be entertained under subsection (1) above if it is made more than six months after the termination of the relevant tenancy.

Date from which interim rent is payable

24B.–(1) The interim rent determined on an application under section 24A(1) of this Act shall be payable from the appropriate date.

(2) If an application under section 24A(1) of this Act is made in a case where the landlord has given a notice under section 25 of this Act, the appropriate date is the earliest date of termination that could have been specified in the landlord's notice.

(3) If an application under section 24A(1) of this Act is made in a case where the tenant has made a request for a new tenancy under section 26 of this Act, the appropriate date is the earliest date that could have been specified in the tenant's request as the date from which the new tenancy is to begin.

Amount of interim rent where new tenancy of whole premises granted and landlord not opposed

24C–(1) This section applies where—

(a) the landlord gave a notice under section 25 of this Act at a time when the tenant was in occupation of the whole of the property comprised in the relevant tenancy for purposes such as are mentioned in section 23(1) of this Act and stated in the notice that he was not opposed to the grant of a new tenancy; or

(b) the tenant made a request for a new tenancy under section 26 of this Act at a time when he was in occupation of the whole of that property for such purposes and the landlord did not give notice under subsection (6) of that section,

and the landlord grants a new tenancy of the whole of the property comprised in the relevant tenancy to the tenant (whether as a result of an order for the grant of a new tenancy or otherwise).

(2) Subject to the following provisions of this section, the rent payable under and at the commencement of the new tenancy shall also be the interim rent.

(3) Subsection (2) above does not apply where—

(a) the landlord or the tenant shows to the satisfaction of the court that the interim rent under that subsection differs substantially from the relevant rent; or

(b) the landlord or the tenant shows to the satisfaction of the court that the terms of the new tenancy differ from the terms of the relevant tenancy to such an extent that the interim rent under that subsection is substantially different from the rent which (in default of such agreement) the court would have determined under section 34 of this

Act to be payable under a tenancy which commenced on the same day as the new tenancy and whose other terms were the same as the relevant tenancy.

(4) In this section "the relevant rent" means the rent which (in default of agreement between the landlord and the tenant) the court would have determined under section 34 of this Act to be payable under the new tenancy if the new tenancy had commenced on the appropriate date (within the meaning of section 24B of this Act).

(5) The interim rent in a case where subsection (2) above does not apply by virtue only of subsection (3)(a) above is the relevant rent.

(6) The interim rent in a case where subsection (2) above does not apply by virtue only of subsection (3)(b) above, or by virtue of subsection (3)(a) and (b) above, is the rent which it is reasonable for the tenant to pay while the relevant tenancy continues by virtue of section 24 of this Act.

(7) In determining the interim rent under subsection (6) above the court shall have regard—

(a) to the rent payable under the terms of the relevant tenancy; and

(b) to the rent payable under any sub-tenancy of part of the property comprised in the relevant tenancy,

but otherwise subsections (1) and (2) of section 34 of this Act shall apply to the determination as they would apply to the determination of a rent under that section if a new tenancy of the whole of the property comprised in the relevant tenancy were granted to the tenant by order of the court and the duration of that new tenancy were the same as the duration of the new tenancy which is actually granted to the tenant.

(8) In this section and section 24D of this Act "the relevant tenancy" has the same meaning as in section 24A of this Act.

Amount of interim rent in any other case

24D–(1) The interim rent in a case where section 24C of this Act does not apply is the rent which it is reasonable for the tenant to pay while the relevant tenancy continues by virtue of section 24 of this Act.

(2) In determining the interim rent under subsection (1) above the court shall have regard—

(a) to the rent payable under the terms of the relevant tenancy; and

(b) to the rent payable under any sub-tenancy of part of the property comprised in the relevant tenancy,

but otherwise subsections (1) and (2) of section 34 of this Act shall apply to the determination as they would apply to the determination of a rent under that section if a new tenancy from year to year of the whole of the property comprised in the relevant tenancy were granted to the tenant by order of the court.

(3) If the court—

(a) has made an order for the grant of a new tenancy and has ordered payment of interim rent in accordance with section 24C of this Act, but

(b) either—

(i) it subsequently revokes under section 36(2) of this Act the order for the grant of a new tenancy; or

(ii) the landlord and tenant agree not to act on the order,

the court on the application of the landlord or the tenant shall determine a new interim rent in accordance with subsections (1) and (2) above without a further application under section 24A(1) of this Act.

Termination of tenancy by the landlord

25–(1) The landlord may terminate a tenancy to which this Part of this Act applies by a notice given to the tenant in the prescribed form specifying the date at which the tenancy is to come to an end (hereinafter referred to as "the date of termination"):

Provided that this subsection has effect subject to **the provisions of section 29B(4) of this Act and** Part IV of this Act as to the interim continuation of tenancies pending the disposal of applications to the court.

(2) Subject to the provisions of the next following subsection, a notice under this section shall not have effect unless it is given not more than twelve nor less than six months before the date of termination specified therein.

(3) In the case of a tenancy which apart from this Act could have been brought to an end by notice to quit given by the landlord—

(a) the date of termination specified in a notice under this section shall not be earlier than the earliest date on which apart from this Part of this Act the tenancy could have been brought to an end by notice to quit given by the landlord on the date of the giving of the notice under this section; and

(b) where apart from this Part of this Act more than six months' notice to quit would have been required to bring the tenancy to an end, the last foregoing subsection shall have effect with the substitution for twelve months of a period six months longer than the length of notice to quit which would have been required as aforesaid.

(4) In the case of any other tenancy, a notice under this section shall not specify a date of termination earlier than the date on which apart from this Part of this Act the tenancy would have come to an end by effluxion of time.

~~(5) A notice under this section shall not have effect unless it requires the tenant, within two months after the giving of the notice, to notify the landlord in writing whether or not, at the date of termination, the tenant will be willing to give up possession of the property comprised in the tenancy.~~

~~(6) A notice under this section shall not have effect unless it states whether the landlord would oppose an application to the court under this Part of this Act for the grant of a new tenancy and, if so, also states on which of the grounds mentioned in section 30 of this Act he would do so.~~

(6) A notice under this section shall not have effect unless it states whether the landlord is opposed to the grant of a new tenancy to the tenant.

(7) A notice under this section which states that the landlord is opposed to the grant of a new tenancy to the tenant shall not have effect unless it also specifies one or more of the grounds specified in section 30(1) of this Act as the ground or grounds for his opposition.

(8) A notice under this section which states that the landlord is not opposed to the grant of a new tenancy to the tenant shall not have effect unless it sets out the landlord's proposals as to—

- **(a) the property to be comprised in the new tenancy (being either the whole or part of the property comprised in the current tenancy);**
- **(b) the rent to be payable under the new tenancy; and**
- **(c) the other terms of the new tenancy.**

Tenant's request for a new tenancy

26–(1) A tenant's request for a new tenancy may be made where the ~~tenancy under which he holds for the time being (hereinafter referred to as "the current tenancy")~~ **current tenancy** is a tenancy granted for a term of years certain exceeding one year, whether or not continued by section 24 of this Act, or granted for a term of years certain and thereafter from year to year.

(2) A tenant's request for a new tenancy shall be for a tenancy beginning with such date, not more than twelve nor less than six months after the making of the request, as may be specified therein;

Provided that the said date shall not be earlier than the date on which apart from this Act the current tenancy would come to an end by effluxion of time or could be brought to an end by notice to quit given by the tenant.

(3) A tenant's request for a new tenancy shall not have effect unless it is made by notice in the prescribed form given to the landlord and sets out the tenant's proposals as to the property to be comprised in the new tenancy (being either the whole or part of the property comprised in the current tenancy), as to the rent to be payable under the new tenancy and as to the other terms of the new tenancy.

(4) A tenant's request for a new tenancy shall not be made if the landlord has already given notice under the last foregoing section to terminate the current tenancy, or if the tenant has already given notice to quit or notice under the next following section; and no such notice shall be given by the landlord or the tenant after the making by the tenant of a request for a new tenancy.

(5) Where the tenant makes a request for a new tenancy in accordance with the foregoing provisions of this section, the current tenancy shall, subject to the provisions of

~~subsection (2) of section thirty six~~ **sections 29B(4) and 36(2)** of this Act and the provisions of Part IV of this Act as to the interim continuation of tenancies, terminate immediately before the date specified in the request for the beginning of the new tenancy.

(6) Within two months of the making of a tenant's request for a new tenancy the landlord may give notice to the tenant that he will oppose an application to the court for the grant of a new tenancy, and any such notice shall state on which of the grounds mentioned in section 30 of this Act the landlord will oppose the application.

Termination by tenant of tenancy for fixed term

27–(1) Where the tenant under a tenancy to which this Part of this Act applies, being a tenancy granted for a term of years certain, gives to the immediate landlord, not later than three months before the date on which apart from this Act the tenancy would come to an end by effluxion of time, a notice in writing that the tenant does not desire the tenancy to be continued, section 24 of this Act shall not have effect in relation to the tenancy, unless the notice is given before the tenant has been in occupation in right of the tenancy for one month.

(1A) Section 24 of this Act shall not have effect in relation to a tenancy for a term of years certain where the tenant is not in occupation of the property comprised in the tenancy at the time when, apart from this Act, the tenancy would come to an end by effluxion of time.

(2) A tenancy granted for a term of years certain which is continuing by virtue of section 24 of this Act **shall not come to an end by reason only of the tenant ceasing to occupy the property comprised in the tenancy but** may be brought to an end on any quarter day by not less than three months' notice in writing given by the tenant to the immediate landlord, whether the notice is given after the date on which apart from this Act the tenancy would have come to an end or before that date, but not before the tenant has been in occupation in right of the tenancy for one month.

(3) Where a tenancy is terminated under subsection (2) above, any rent payable in respect of a period which begins before, and ends after, the tenancy is terminated shall be apportioned, and any rent paid by the tenant in excess of the amount apportioned to the period before termination shall be recoverable by him.

Renewal of tenancies by agreement

28 Where the landlord and tenant agree for the grant to the tenant of a future tenancy of the holding, or of the holding with other land, on terms and from a date specified in the agreement, the current tenancy shall continue until that date but no longer, and shall not be a tenancy to which this Part of this Act applies.

~~*Application to court for new tenancies*~~

~~**Order by court for grant of a new tenancy**~~

~~29. (1) Subject to the provisions of this Act, on an application under subsection (1) of section 24 of this Act for a new tenancy the court shall make an order for the grant of a tenancy comprising such property, at such rent and on such other terms, as are hereinafter provided.~~

~~(2) Where such an application is made in consequence of a notice given by the landlord under section 25 of this Act, it shall not be entertained unless the tenant has duly notified the landlord that he will not be willing at the date of termination to give up possession of the property comprised in the tenancy.~~

~~(3) No application under subsection (1) of section 24 of this Act shall be entertained unless it is made not less than two nor more than four months after the giving of the landlord's notice under section 25 of this Act or, as the case may be, after the making of the tenant's request for a new tenancy.~~

Applications to court

Order by court for grant of new tenancy or termination of current tenancy

29–(1) Subject to the provisions of this Act, on an application under section 24(1) of this Act, the court shall make an order for the grant of a new tenancy and accordingly for the termination of the current tenancy immediately before the commencement of the new tenancy.

(2) Subject to the following provisions of this Act, a landlord may apply to the court for an order for the termination of a tenancy to which this Part of this Act applies without the grant of a new tenancy—

(a) if he has given notice under section 25 of this Act that he is opposed to the grant of a new tenancy to the tenant; or

(b) if the tenant has made a request for a new tenancy in accordance with section 26 of this Act and the landlord has given notice under subsection (6) of that section.

(3) The landlord may not make an application under subsection (2) above if either the tenant or the landlord has made an application under section 24(1) of this Act.

(4) Subject to the provisions of this Act, where the landlord makes an application under subsection (2) above—

(a) if he establishes, to the satisfaction of the court, any of the grounds on which he is entitled to make the application in accordance with section 30 of this Act, the court shall make an order for the termination of the current tenancy in accordance with section 64 of this Act without the grant of a new tenancy; and

(b) if not, it shall make an order for the grant of a new tenancy and accordingly for the termination of the current tenancy immediately before the commencement of the new tenancy.

(5) The court shall dismiss an application by the landlord under section 24(1) of this Act if the tenant informs the court that he does not want a new tenancy.

(6) The landlord may not withdraw an application under subsection (2) above unless the tenant consents to its withdrawal.

Time limits for applications to court

29A.–(1) Subject to section 29B of this Act, the court shall not entertain an application—

(a) by the tenant or the landlord under section 24(1) of this Act; or

(b) by the landlord under section 29(2) of this Act,

if it is made after the end of the statutory period.

(2) In this section and section 29B of this Act "the statutory period" means a period ending—

(a) where the landlord gave a notice under section 25 of this Act, on the date specified in his notice; and

(b) where the tenant made a request for a new tenancy under section 26 of this Act, immediately before the date specified in his request.

(3) Where the tenant has made a request for a new tenancy under section 26 of this Act, the court shall not entertain an application under section 24(1) of this Act which is made before the end of the period of two months beginning with the date of the making of the request, unless the application is made after the landlord has given a notice under section 26(6) of this Act.

Agreements extending time limits

29B.–(1) After the landlord has given a notice under section 25 of this Act, or the tenant has made a request under section 26 of this Act, but before the end of the statutory period, the landlord and tenant may agree that an application such as is mentioned in section 29A(1) of this Act may be made before the end of a period specified in the agreement which will expire after the end of the statutory period.

(2) The landlord and tenant may from time to time by agreement further extend the period for making such an application, but any such agreement must be made before the end of the period specified in the current agreement.

(3) Where an agreement is made under this section, the court may entertain an application such as is mentioned in section 29A(1) of this Act if it is made before the end of the period specified in the agreement.

(4) Where an agreement is made under this section, or two or more agreements are made under this section, the landlord's notice under section 25 of this Act or tenant's request under section 26 of this Act shall be treated as terminating the tenancy at the end of the period specified in the agreement or, as the case may be, at the end of the period specified in the last of those agreements.

Opposition by landlord to application for new tenancy

30–(1) The grounds on which a landlord may oppose an application under ~~subsection (1) of section twenty-four of this Act~~ **section 24(1) of this Act, or make an application under section 29(2) of this Act**, are such of the following grounds as may be stated in the landlord's notice under section 25 of this Act or, as the case may be, under subsection (6) of section 26 thereof, that is to say—

(a) where under the current tenancy the tenant has any obligations as respects the repair and maintenance of the holding, that the tenant ought not to be granted a new tenancy in view of the state of repair of the holding, being a state resulting from the tenant's failure to comply with the said obligations;

(b) that the tenant ought not to be granted a new tenancy in view of his persistent delay in paying rent which has become due;

(c) that the tenant ought not to be granted a new tenancy in view of other substantial breaches by him of his obligations under the current tenancy, or for any other reason connected with the tenant's use or management of the holding;

(d) that the landlord has offered and is willing to provide or secure the provision of alternative accommodation for the tenant, that the terms on which the alternative accommodation is available are reasonable having regard to the terms of the current tenancy and to all other relevant circumstances, and that the accommodation and the time at which it will be available are suitable for the tenant's requirements (including the requirement to preserve goodwill) having regard to the nature and class of his business and to the situation and extent of, and facilities afforded by, the holding;

(e) where the current tenancy was created by the sub-letting of part only of the property comprised in a superior tenancy and the landlord is the owner of an interest in reversion expectant on the termination of that superior tenancy, that the aggregate of the rents reasonably obtainable on separate lettings of the holding and the remainder of that property would be substantially less than the rent reasonably obtainable on a letting of that property as a whole, that on the termination of the current tenancy the landlord requires possession of the holding for the purpose of letting or otherwise disposing of the said property as a whole, and that in view thereof the tenant ought not to be granted a new tenancy;

(f) that on the termination of the current tenancy the landlord intends to demolish or reconstruct the premises comprised in the holding or a substantial part of those premises or to carry out substantial work of construction on the holding or part thereof and that he could not reasonably do so without obtaining possession of the holding;

(g) subject as hereinafter provided, that on the termination of the current tenancy the landlord intends to occupy the holding for the purposes, or partly for the purposes, of a business to be carried on by him therein, or as his residence.

(1A) Where the landlord has a controlling interest in a company, the reference in subsection (1)(g) above to the landlord shall be construed as a reference to the landlord or that company.

(1B) Subject to subsection (2A) below, where the landlord is a company and a person has a controlling interest in the company, the reference in subsection (1)(g) above to the landlord shall be construed as a reference to the landlord or that person.

(2) The landlord shall not be entitled to oppose an application **under section 24(1) of this Act, or make an application under section 29(2) of this Act**, on the ground specified in paragraph (g) of the last foregoing subsection if the interest of the landlord, or an interest which has merged in that interest and but for the merger would be the interest of the landlord, was purchased or created after the beginning

of the period of five years which ends with the termination of the current tenancy, and at all times since the purchase or creation thereof the holding has been comprised in a tenancy or successive tenancies of the description specified in subsection (1) of section 23 of this Act.

(2A) Subsection (1B) above shall not apply if the controlling interest was acquired after the beginning of the period of five years which ends with the termination of the current tenancy, and at all times since the acquisition of the controlling interest the holding has been comprised in a tenancy or successive tenancies of the description specified in section 23(1) of this Act.

~~(3) Where the landlord has a controlling interest in a company any business to be carried on by the company shall be treated for the purposes of subsection (1) (g) of this section as a business to be carried on by him.~~

~~For the purposes of this subsection, a person has a controlling interest in a company if and only if either –~~

~~(a) he is a member of it and able, without the consent of any other person, to appoint or remove the holders of at least a majority of the directorships; or~~

~~(b) he holds more than one half of its equity share capital, there being disregarded any shares held by him in a fiduciary capacity or as nominee for another person;~~

~~and in this subsection "company" and "share" have the meanings assigned to them by section 455(1) of the Companies Act 1948 and "equity share capital" the meaning assigned to it by section 154(5) of that Act.~~

Dismissal of application for new tenancy where landlord successfully opposes

31–(1) If the landlord opposes an application under subsection (1) of section 24 of this Act on grounds on which he is entitled to oppose it in accordance with the last foregoing section and establishes any of those grounds to the satisfaction of the court, the court shall not make an order for the grant of a new tenancy.

(2) Where the landlord opposes an application under section 24(1) of this Act, or makes an application under section 29(2) of this Act, on one or more of the grounds specified in section 30(1)(d) to (f) of this Act but establishes none of those grounds, and none of the other grounds specified in section 30(1) of this Act, to the satisfaction of the court, then if the court would have been satisfied on any of the grounds specified in section 30(1)(d) to (f) of this Act ~~Where in a case not falling within the last foregoing subsection the landlord opposes an application under the said subsection (1) on one or more of the grounds specified in paragraphs (d), (e) and (f) of subsection (1) of the last foregoing section but establishes none of those grounds to the satisfaction of the court; then if the court would have been satisfied of any of those grounds~~ if the date of termination specified in the landlord's notice or, as the case may be, the date specified in the tenant's request for a new tenancy as the date from which the new tenancy is to begin, had been such later date as the court may determine, being a date not more than one year later than the date so specified,-

(a) the court shall make a declaration to that effect, stating of which of the said grounds the court would have been satisfied as aforesaid and specifying the date determined by the court as aforesaid, but shall not make an order for the grant of a new tenancy;

(b) if, within fourteen days after the making of the declaration, the tenant so requires the court shall make an order substituting the said date for the date specified in the said landlord's notice or tenant's request, and thereupon that notice or request shall have effect accordingly.

Grant of new tenancy in some cases where section 30(1)(f) applies

31A–(1) Where the landlord opposes an application under section 24(1) of this Act on the ground specified in paragraph (f) of section 30(1) of this Act, **or makes an application under section 29(2) of this Act on that ground,** the court shall not hold that the landlord could not reasonably carry out the demolition, reconstruction or work of construction intended without obtaining possession of the holding if—

(a) the tenant agrees to the inclusion in the terms of the new tenancy of terms giving the landlord access and other facilities for carrying out the work intended and, given that access and those facilities, the landlord could reasonably carry out the work without obtaining possession of the holding and without interfering to a substantial extent or for a substantial time with the use of the holding for the purposes of the business carried on by the tenant; or

(b) the tenant is willing to accept a tenancy of an economically separable part of the holding and either paragraph (a) of this section is satisfied with respect to that part or possession of the remainder of the holding would be reasonably sufficient to enable the landlord to carry out the intended work.

(2) For the purposes of subsection (1) (b) of this section a part of a holding shall be deemed to be an economically separate part if, and only if, the aggregate of the rents which, after the completion' of the intended work, would be reasonably obtainable on separate lettings of that part and the remainder of the premises affected by or resulting from the work would not be substantially less than the rent which would then be reasonably obtainable on a letting of those premises as a whole.

Property to be comprised in new tenancy

32–(1) Subject to the following provisions of this section, an order under section 29 of this Act for the grant of a new tenancy shall be an order for the grant of a new tenancy of the holding; and in the absence of agreement between the landlord and the tenant as to the property which constitutes the holding the court shall in the order designate that property by reference to the circumstances existing at the date of the order.

(1A) Where the court, by virtue of paragraph (b) of section 31A(1) of this Act, makes an order under section 29 of this Act for the grant of a new tenancy in a case where the tenant is willing to accept a tenancy of part of the holding, the order shall be an order for the grant of a new tenancy of that part only.

(2) The foregoing provisions of this section shall not apply in a case where the property comprised in the current tenancy includes other property besides the holding and the landlord requires any new tenancy ordered to be granted under section 29 of this Act to be a tenancy of the whole of the property comprised in the current tenancy; but in any such case—

(a) any order under the said section 29 for the grant of a new tenancy shall be an order for the grant of a new tenancy of the whole of the property comprised in the current tenancy, and

(b) references in the following provisions of this Part of this Act to the holding shall be construed as references to the whole of that property.

(3) Where the current tenancy includes rights enjoyed by the tenant in connection with the holding, those rights shall be included in a tenancy ordered to be granted under section 29 of this Act, except as otherwise agreed between the landlord and the tenant or, in default of such agreement, determined by the court.

Duration of new tenancy

33. Where on an application under this Part of this Act the court makes an order for the grant of a new tenancy, the new tenancy shall be such tenancy as may be agreed between the landlord and the tenant, or, in default of such an agreement, shall be such a tenancy as may be determined by the court to be reasonable in all the circumstances, being, if it is a tenancy for a term of years certain, a tenancy for a term not exceeding ~~fourteen~~ **fifteen** years, and shall begin on the coming to an end of the current tenancy.

Rent under new tenancy

34–(1) The rent payable under a tenancy granted by order of the court under this Part of this Act shall be such as may be agreed between the landlord and the tenant or as, in default of such agreement, may be determined by the court to be that at which, having regard to the terms of the tenancy (other than those relating to rent), the holding might reasonably be expected to be let in the open market by a willing lessor, there being disregarded—

(a) any effect on rent of the fact that the tenant has or his predecessors in title have been in occupation of the holding,

(b) any goodwill attached to the holding by reason of the carrying on thereat of the business of the tenant (whether by him or by a predecessor of his in that business),

(c) any effect on rent of an improvement to which this paragraph applies,

(d) in the case of a holding comprising licensed premises, any addition to its value attributable to the licence, if it appears to the court that having regard to the terms of the current tenancy and any other relevant circumstances the benefit of the licence belongs to the tenant.

(2) Paragraph (c) of the foregoing subsection applies to any improvement carried out by a person who at the time it was carried out was the tenant, but only if it was carried out otherwise than in pursuance of an obligation to his immediate landlord, and either it was carried out during the current tenancy or the following conditions are satisfied, that is to say—

(a) that it was completed not more than twenty-one years before the application ~~for the new tenancy~~ **to the court** was made; and

(b) that the holding or any part of it affected by the improvement has at all times since the completion of the improvement been comprised in tenancies of the description specified in section 23(1) of this Act; and

(c) that at the termination of each of those tenancies the tenant did not quit.

(2A) If this Part of this Act applies by virtue of section 23(1A) of this Act, the reference in subsection (1)(d) above to the tenant shall be construed as including—

(a) a company in which the tenant has a controlling interest, or

(b) where the tenant is a company, a person with a controlling interest in the company.

(3) Where the rent is determined by the court the court may, if it thinks fit, further determine that the terms of the tenancy shall include such provision for varying the rent as may be specified in the determination.

(4) It is hereby declared that the matters which are to be taken into account by the court in determining the rent include any effect on rent of the operation of the provisions of the Landlord and Tenant (Covenants) Act 1995.

Other terms of new tenancy

35–(1) The terms of a tenancy granted by order of the court under this Part of this Act (other than terms as to the duration thereof and as to the rent payable thereunder), **including, where different persons own interests which fulfil the conditions specified in section 44(1) of this Act in different parts of it, terms as to the apportionment of the rent,** shall be such as may be agreed between the landlord and the tenant or as, in default of such agreement, may be determined by the court; and in determining those terms the court shall have regard to the terms of the current tenancy and to all relevant circumstances.

(2) In subsection (1) of this section the reference to all relevant circumstances includes (without prejudice to the generality of that reference) a reference to the operation of the provisions of the Landlord and Tenant (Covenants) Act 1995.

Carrying out of order for new tenancy

36–(1) Where under this Part of this Act the court makes an order for the grant of a new tenancy, then, unless the order is revoked under the next following subsection or the landlord and the tenant agree not to act upon the order, the landlord shall be bound to execute or make in favour of the tenant, and the tenant shall be bound to accept, a lease or agreement for a tenancy of the holding embodying the terms agreed between the landlord and the tenant or determined by the court in accordance with the foregoing provisions of this Part of this Act; and
where the landlord executes or makes such a lease or agreement the tenant shall be bound, if so required by the landlord, to execute a counterpart or duplicate thereof.

(2) If the tenant, within fourteen days after the making of an order under this Part of this Act for the grant of a new tenancy, applies to the court for the revocation of the order the court shall revoke the order; and where the order is so revoked, then, if it is so agreed between the landlord and the tenant or determined by the court, the current tenancy shall continue, beyond the date at which it would have come to an end apart from this subsection, for such period as may be so agreed or determined to be

necessary to afford to the landlord a reasonable opportunity for reletting or otherwise disposing of the premises which would have been comprised in the new tenancy; and while the current tenancy continues by virtue of this subsection it shall not be a tenancy to which this Part of this Act applies.

(3) Where an order is revoked under the last foregoing subsection any provision thereof as to payment of costs shall not cease to have effect by reason only of the revocation; but the court may, if it thinks fit, revoke or vary any such provision or, where no costs have been awarded in the proceedings for the revoked order, award such costs.

(4) A lease executed or agreement made under this section, in a case where the interest of the lessor is subject to a mortgage, shall be deemed to be one authorised by section 99 of the Law of Property Act 1925 (which confers certain powers of leasing on mortgagors in possession), and subsection (13) of that section (which allows those powers to be restricted or excluded by agreement) shall not have effect in relation to such a lease or agreement.

Compensation where order for new tenancy precluded on certain grounds

~~37 (1) Where on the making of an application under section 24 of this Act the court is precluded (whether by subsection (1) or subsection (2) of section 31 of this Act) from making an order for the grant of a new tenancy by reason of any of the grounds specified in paragraphs (e), (f) and (g) of subsection (1) of section 30 of this Act and not of any grounds specified in any other paragraph of that subsection, or where no other ground is specified in the landlord's notice under section 25 of this Act or, as the case may be, under section 26(6) thereof, than those specified in the said paragraphs (e), (f) and (g) and either no application under the said section 24 is made or such an application is withdrawn, then, subject to the provisions of this Act, the tenant shall be entitled on quitting the holding to recover from the landlord by way of compensation an amount determined in accordance with the following provisions of this section.~~

37–(1) Subject to the provisions of this Act, in a case specified in subsection (1A), (1B) or (1C) below (a "compensation case") the tenant shall be entitled on quitting the holding to recover from the landlord by way of compensation an amount determined in accordance with this section.

(1A) The first compensation case is where on the making of an application by the tenant under section 24(1) of this Act the court is precluded (whether by subsection (1) or subsection (2) of section 31 of this Act) from making an order for the grant of a new tenancy by reason of any of the grounds specified in paragraphs (e), (f) and (g) of section 30(1) of this Act (the "compensation grounds") and not of any grounds specified in any other paragraph of section 30(1).

(1B) The second compensation case is where on the making of an application under section 29(2) of this Act the court is precluded (whether by section 29(4)(a) or section 31(2) of this Act) from making an order for the grant of a new tenancy by reason of any of the compensation grounds and not of any other grounds specified in section 30(1) of this Act.

(1C) The third compensation case is where—

(a) the landlord's notice under section 25 of this Act or, as the case may be, under section 26(6)

of this Act, states his opposition to the grant of a new tenancy on any of the compensation grounds and not on any other grounds specified in section 30(1) of this Act; and

(b) either—

(i) no application is made by the tenant under section 24(1) of this Act or by the landlord under section 29(2) of this Act; or

(ii) such an application is made but is subsequently withdrawn.

(2) Subject to ~~subsections (5A) to (5E) of this section the said amount~~ **the following provisions of this section, compensation under this section** shall be as follows, that is to say—

(a) where the conditions specified in the next following subsection are satisfied **in relation to the whole of the holding** it shall be the product of the appropriate multiplier and twice the rateable value of the holding,

(b) in any other case it shall be the product of the appropriate multiplier and the rateable value of the holding.

(3) The said conditions are—

(a) that, during the whole of the fourteen years immediately preceding the termination of the current tenancy, premises being or comprised in the holding have been occupied for the purposes of a business carried on by the occupier or for those and other purposes;

(b) that, if during those fourteen years there was a change in the occupier of the premises, the person who was the occupier immediately after the change was the successor to the business carried on by the person who was the occupier immediately before the change.

(3A) If the conditions specified in subsection (3) above are satisfied in relation to part of the holding but not in relation to the other part, the amount of compensation shall be the aggregate of sums calculated separately as compensation in respect of each part, and accordingly, for the purpose of calculating compensation in respect of a part any reference in this section to the holding shall be construed as a reference to that part.

(3B) Where section 44(1A) of this Act applies, the compensation shall be determined separately for each part and compensation determined for any part shall be recoverable only from the person who is the owner of an interest in that part which fulfils the conditions specified in section 44(1) of this Act.

(4) Where the court is precluded from making an order for the grant of a new tenancy under this Part of this Act in ~~the circumstances mentioned in subsection (1) of this section~~ **a compensation case**, the court shall on the application of the tenant certify that fact.

(5) For the purposes of subsection (2) of this section the rateable value of the holding shall be determined as follows:-

(a where in the valuation list in force at the date on which the landlord's notice under section 25 or, as the case may be, subsection (6) of section 26 of this Act is given a value is then shown as the annual value (as hereinafter defined) of the holding, the rateable value of the holding shall be taken to be that value;

(b) where no such value is so shown with respect to the holding but such a value or such values is or are so shown with respect to premises comprised in or comprising the holding or part of it, the rateable value of the holding shall be taken to be such value as is found by a proper apportionment or aggregation of the value or values so shown;

(c) where the rateable value of the holding cannot be ascertained in accordance with the foregoing paragraphs of this subsection, it shall be taken to be the value which, apart from any exemption from assessment to rates, would on a proper assessment be the value to be entered in the said valuation list as the annual value of the holding;

and any dispute arising, whether in proceedings before the court or otherwise, as to the determination for those purposes of the rateable value of the holding shall be referred to the Commissioners of Inland Revenue for decision by the valuation officer.

An appeal shall lie to the Lands Tribunal from any decision of a valuation officer under this subsection, but subject thereto any such decision shall be final.

(5A) If part of the holding is domestic property, as defined in section 66 of the Local Government Finance Act 1988—

(a) the domestic property shall be disregarded in determining the rateable value of the holding under subsection (5) of this section; and

(b) if, on the date specified in subsection (5) (a) of this section, the tenant occupied the whole or any part of the domestic property, the amount of compensation to which he is entitled under subsection (1) of this section shall be increased by the addition of a sum equal to his reasonable expenses in removing from the domestic property.

(5B) Any question as to the amount of the sum referred to in paragraph (b) of subsection (5A) of this section shall be determined by agreement between the landlord and the tenant or, in default of agreement, by the court.

(5C) If the whole of the holding is domestic property, as defined in section 66 of the Local Government Finance Act 1988, for the purposes of subsection (2) of this section the rateable value of the holding shall be taken to be an amount equal to the rent at which it is estimated the holding might reasonably be expected to let from year to year if the tenant undertook to pay all usual tenant's rates and taxes and to bear the cost of the repairs and insurance and the other expenses (if any) necessary to maintain the holding in a state to command that rent.

(5D) The following provisions shall have effect as regards a determination of an amount mentioned in subsection (5C) of this section—

(a) the date by reference to which such a determination is to be made is the date on which the landlord's notice under section 25 or, as the case may be, subsection (6) of section 26 of this Act is given;

(b) any dispute arising, whether in proceedings before the court or otherwise, as to such a determination shall be referred to the Commissioners of Inland Revenue for decision by a valuation officer;

(c) an appeal shall lie to the Lands Tribunal from such a decision, but subject to that, such a decision shall be final.

(5E) Any deduction made under paragraph 2A of Schedule 6 to the Local Government Finance Act 1988 (deduction from valuation of hereditaments used for breeding horses etc.) shall be disregarded, to the extent that it relates to the holding, in determining the rateable value of the holding under subsection (5) of this section.

(6) The Commissioners of Inland Revenue may by statutory instrument make rules prescribing the procedure in connection with references under this section.

(7) In this section—

the reference to the termination of the current tenancy is a reference to the date of termination specified in the landlord's notice under section 25 of this Act or, as the case may be, the date specified in the tenant's request for a new tenancy as the date from which the new tenancy is to begin;

the expression "annual value" means rateable value except that where the rateable value differs from the net annual value the said expression means net annual value;

the expression "valuation officer" means any officer of the Commissioners of Inland Revenue for the time being authorised by a certificate of the Commissioners to act in relation to a valuation list.

(8) In subsection (2) of this section "the appropriate multiplier" means such multiplier as the Secretary of State may by order made by statutory instrument prescribe and different multipliers may be so prescribed in relation to different cases.

(9) A statutory instrument containing an order under subsection (8) of this section shall be subject to annulment in pursuance of a resolution of either House of Parliament.

Compensation for possession obtained by misrepresentation

37A–(1) Where the court—

(a) makes an order for the termination of the current tenancy but does not make an order for the grant of a new tenancy, or

(b) refuses an order for the grant of a new tenancy,

and it is subsequently made to appear to the court that the order was obtained, or the court was induced to refuse the grant, by misrepresentation or the concealment of material facts, the court may order the landlord to pay to the tenant such sum as appears sufficient as compensation for damage or loss sustained by the tenant as the result of the order or refusal.

(2) Where—

(a) the tenant has quit the holding—

(i) after making but withdrawing an application under section 24(1) of this Act; or

(ii) without making such an application; and

(b) it is made to appear to the court that he did so by reason of misrepresentation or the concealment of material facts,

the court may order the landlord to pay to the tenant such sum as appears sufficient as compensation for damage or loss sustained by the tenant as the result of quitting the holding.

Restriction on agreements excluding provisions of Part II

38–(1) Any agreement relating to a tenancy to which this Part of this Act applies (whether contained in the instrument creating the tenancy or not) shall be void (except as provided by ~~subsection (4) of this section~~ **section 38A of this Act**) in so far as it purports to preclude the tenant from making an application or request under this Part of this Act or provides for the termination or the surrender of the tenancy in the event of his making such an application or request or for the imposition of any penalty or disability on the tenant in that event.

(2) Where—

(a) during the whole of the five years immediately preceding the date on which the tenant under a tenancy to which this Part of this Act applies is to quit the holding, premises being or comprised in the holding have been occupied for the purposes of a business carried on by the occupier or for those and other purposes, and

(b) if during those five years there was a change in the occupier of the premises, the person who was the occupier immediately after the change was the successor to the business carried on by the person who was the occupier immediately before the change,

any agreement (whether contained in the instrument creating the tenancy or not and whether made before or after the termination of that tenancy) which purports to exclude or reduce compensation under ~~the last foregoing section~~ **section 37 of this Act** shall to that extent be void, so however that this subsection shall not affect any agreement as to the amount of any such compensation which is made after the right to compensation has accrued.

(3) In a case not falling within the last foregoing subsection the right to compensation conferred ~~by the last foregoing section~~ **section 37 of this Act** may be excluded or modified by agreement.

~~(4) The court may—~~

~~(a) on the joint application of the persons who will be the landlord and the tenant in relation to a tenancy to be granted for a term of years certain which will be a tenancy to which this Part of this Act applies, authorise an agreement excluding in relation to that tenancy the provisions of sections 24 to 28 of this Act; and~~

~~(b) on the joint application of the persons who are the landlord and the tenant in relation to a tenancy to which this Part of this Act applies, authorise an agreement for the surrender of the~~

~~tenancy on such date or in such circumstances as may be specified in the agreement and on such terms (if any) as may be so specified;~~

~~if the agreement is contained in or endorsed on the instrument creating the tenancy or such other instrument as the court may specify; and an agreement contained in or endorsed on an instrument in pursuance of an authorisation given under this subsection shall be valid notwithstanding anything in the preceding provisions of this section.~~

Agreements to exclude provisions of Part II

38A–(1) The persons who will be the landlord and the tenant in relation to a tenancy to be granted for a term of years certain which will be a tenancy to which this Part of this Act applies may agree that the provisions of sections 24 to 28 of this Act shall be excluded in relation to that tenancy.

(2) The persons who are the landlord and the tenant in relation to a tenancy to which this Part of this Act applies may agree that the tenancy shall be surrendered on such date or in such circumstances as may be specified in the agreement and on such terms (if any) as may be so specified.

(3) An agreement under subsection (1) above shall be void unless—

(a) the landlord has served on the tenant a notice in the form, or substantially in the form, set out in Schedule 1 to the Regulatory Reform (Business Tenancies) (England and Wales) Order 2003 ("the 2003 Order"); and

(b) the requirements specified in Schedule 2 to that Order are met.

(4) An agreement under subsection (2) above shall be void unless—

(a) the landlord has served on the tenant a notice in the form, or substantially in the form, set out in Schedule 3 to the 2003 Order; and

(b) the requirements specified in Schedule 4 to that Order are met.

General and supplementary provisions

Saving for compulsory acquisitions

39–(1) [Repealed]

(2) If the amount of the compensation which would have been payable under section 37 of this Act if the tenancy had come to an end in circumstances giving rise to compensation under that section and the date at which the acquiring authority obtained possession had been the termination of the current tenancy exceeds the amount of the compensation payable under section 121 of the Lands Clauses Consolidation Act 1845, or section 20 of the Compulsory Purchase Act 1965, in the case of a tenancy to which this Part of this Act applies, that compensation shall be increased by the amount of the excess.

(3) Nothing in section 24 of this Act shall affect the operation of the said section 121.

Duty of tenants and landlords of business premises to give information to each other

~~**40.** (1) Where any person having an interest in any business premises, being an interest in reversion expectant (whether immediately or not) on a tenancy of those premises, serves on the tenant a notice in the prescribed form requiring him to do so, it shall be the duty of the tenant to notify that person in writing within one month of the service of the notice—~~

~~(a) whether he occupies the premises or any part thereof wholly or partly for the purposes of a business carried on him, and~~

~~(b) whether his tenancy has effect subject to any sub-tenancy on which his tenancy is immediately expectant and, if so, what premises are comprised in the sub-tenancy, for what term it has effect (or, if it is terminable by notice, by what notice it can be terminated), what is the rent payable thereunder, who is the sub-tenant, and (to the best of his knowledge and belief) whether the sub-tenant is in occupation of the premises or of part of the premises comprised in the sub-tenancy and, if not, what is the sub-tenant's address.~~

~~(2) Where the tenant of any business premises, being a tenant under such a tenancy as is mentioned in subsection (1) of section 26 of this Act, service on any persons mentioned in tire next following subsection a notice in the prescribed form requiring him to do so, it shall be the duty of that person to notify the tenant in writing within one month after the service of the notice-~~

~~(a) whether he is the owner of the fee simple in respect of those premises or any part thereof or the mortgagee in possession of such an owner and, if not,~~

~~(b) (to the best of his knowledge and belief) the name and address of the person who is his or, as the case may be, his mortgagor's immediate landlord in respect of those premises or of the part in respect of which he or his mortgagor is not the owner in fee simple, for what term his or his mortgagor's tenancy thereof has effect and what is the earliest date (if any) at which that tenancy is terminable by notice to quit given by the landlord.~~

~~(3) The persons referred to in the last foregoing subsection are, in relation to the tenant of any business premises—~~

~~(a) any person having an interest in the premises, being an interest in reversion expectant (whether immediately or not) on the tenant's, and~~

~~(b) any person being a mortgagee in possession in respect of such an interest in reversion as is mentioned in paragraph (a) of this subsection;~~

~~and the information which any such person as is mentioned in paragraph (a) of this subsection is required to give under the last foregoing subsection shall include information whether there is a mortgagee in possession of his interest in the premises and, if so, what is the name and address of the mortgagee.~~

~~(4) The foregoing provisions of this section shall not apply to a notice served by or on the tenant more than two years before the date on which apart from this Act his tenancy would come to an end by effluxion of time or could be brought to an end by notice to quit given by the landlord.~~

(5) In this section

the expression "business premises" means premises used wholly or partly for the purposes of a business;

~~the expression "mortgagee in possession" includes a receiver appointed by the mortgagee or by the court who is in receipt of the rents and profits, and the expression "his mortgagor" shall be construed accordingly;~~

~~the expression "sub-tenant" includes a person retaining possession of any premises by virtue of the Rent Act 1977 after the coming to an end of a sub-tenancy, and the expression "sub-tenancy" includes a right so to retain possession.~~

40–(1) Where a person who is an owner of an interest in reversion expectant (whether immediately or not) on a tenancy of any business premises has served on the tenant a notice in the prescribed form requiring him to do so, it shall be the duty of the tenant to give the appropriate person in writing the information specified in subsection (2) below.

(2) That information is—

(a) whether the tenant occupies the premises or any part of them wholly or partly for the purposes of a business carried on by him;

(b) whether his tenancy has effect subject to any sub-tenancy on which his tenancy is immediately expectant and, if so—

(i) what premises are comprised in the sub-tenancy;

(ii) for what term it has effect (or, if it is terminable by notice, by what notice it can be terminated);

(iii) what is the rent payable under it;

(iv) who is the sub-tenant;

(v) (to the best of his knowledge and belief) whether the sub-tenant is in occupation of the premises or of part of the premises comprised in the sub-tenancy and, if not, what is the sub-tenant's address;

(vi) whether an agreement is in force excluding in relation to the sub-tenancy the provisions of sections 24 to 28 of this Act; and

(vii) whether a notice has been given under section 25 or 26(6) of this Act, or a request has been made under section 26 of this Act, in relation to the sub-tenancy and, if so, details of the notice or request; and

(c) (to the best of his knowledge and belief) the name and address of any other person who owns an interest in reversion in any part of the premises.

(3) Where the tenant of any business premises who is a tenant under such a tenancy as is mentioned in section 26(1) of this Act has served on a reversioner or a reversioner's mortgagee in possession a notice in the prescribed form requiring him to do so, it shall be the duty of the person on whom the notice is served to give the appropriate person in writing the information specified in subsection (4) below.

(4) That information is—

(a) whether he is the owner of the fee simple in respect of the premises or any part of them or the mortgagee in possession of such an owner,

(b) if he is not, then (to the best of his knowledge and belief)—

(i) the name and address of the person who is his or, as the case may be, his mortgagor's immediate landlord in respect of those premises or of the part in respect of which he or his mortgagor is not the owner in fee simple;

(ii) for what term his or his mortgagor's tenancy has effect and what is the earliest date (if any) at which that tenancy is terminable by notice to quit given by the landlord; and

(iii) whether a notice has been given under section 25 or 26(6) of this Act, or a request has been made under section 26 of this Act, in relation to the tenancy and, if so, details of the notice or request;

(c) (to the best of his knowledge and belief) the name and address of any other person who owns an interest in reversion in any part of the premises; and

(d) if he is a reversioner, whether there is a mortgagee in possession of his interest in the premises and, if so, (to the best of his knowledge and belief) what is the name and address of the mortgagee.

(5) A duty imposed on a person by this section is a duty—

(a) to give the information concerned within the period of one month beginning with the date of service of the notice; and

(b) if within the period of six months beginning with the date of service of the notice that person becomes aware that any information which has been given in pursuance of the notice is not, or is no longer, correct, to give the appropriate person correct information within the period of one month beginning with the date on which he becomes aware.

(6) This section shall not apply to a notice served by or on the tenant more than two years before the date on which apart from this Act his tenancy would come to an end by effluxion of time or could be brought to an end by notice to quit given by the landlord.

(7) Except as provided by section 40A of this Act, the appropriate person for the purposes of this section and section 40A(1) of this Act is the person who served the notice under subsection (1) or (3) above.

(8) In this section—

"business premises" means premises used wholly or partly for the purposes of a business;

"mortgagee in possession" includes a receiver appointed by the mortgagee or by the court who is in receipt of the rents and profits, and "his mortgagor" shall be construed accordingly;

"reversioner" means any person having an interest in the premises, being an interest in reversion expectant (whether immediately or not) on the tenancy;

"reversioner's mortgagee in possession" means any person being a mortgagee in possession in respect of such an interest; and

"sub-tenant" includes a person retaining possession of any premises by virtue of the Rent (Agriculture) Act 1976 or the Rent Act 1977 after the coming to an end of a sub-tenancy, and "sub-tenancy" includes a right so to retain possession.

Duties in transfer cases

40A–(1) If a person on whom a notice under section 40(1) or (3) of this Act has been served has transferred his interest in the premises or any part of them to some other person and gives the appropriate person notice in writing—

- (a) of the transfer of his interest; and
- (b) of the name and address of the person to whom he transferred it,

on giving the notice he ceases in relation to the premises or (as the case may be) to that part to be under any duty imposed by section 40 of this Act.

(2) If—

- (a) the person who served the notice under section 40(1) or (3) of this Act ("the transferor") has transferred his interest in the premises to some other person ("the transferee"); and
- (b) the transferor or the transferee has given the person required to give the information notice in writing—
  - (i) of the transfer; and
  - (ii) of the transferee's name and address,

the appropriate person for the purposes of section 40 of this Act and subsection (1) above is the transferee.

(3) If—

- (a) a transfer such as is mentioned in paragraph (a) of subsection (2) above has taken place; but
- (b) neither the transferor nor the transferee has given a notice such as is mentioned in paragraph (b) of that subsection,

any duty imposed by section 40 of this Act may be performed by giving the information either to the transferor or to the transferee.

Proceedings for breach of duties to give information

40B A claim that a person has broken any duty imposed by section 40 of this Act may be made the subject of civil proceedings for breach of statutory duty; and in any such proceedings a court may order that person to comply with that duty and may make an award of damages.

Trusts

41–(1) Where a tenancy is held on trust, occupation by all or any of the beneficiaries under the trust, and the carrying on of a business by all or any of the beneficiaries, shall be treated for the purposes of section 23 of this Act as equivalent to occupation or the carrying on of a business by the tenant; and in relation to a tenancy to which this Part of this Act applies by virtue of the foregoing provisions of this subsection—

(a) references (however expressed) in this Part of this Act and in the Ninth Schedule to this Act to the business of, or to carrying on of business, use, occupation or enjoyment by, the tenant shall be construed as including references to the business of, or to carrying on of business, use, occupation or enjoyment by, the beneficiaries or beneficiary;

(b) the reference in paragraph (d) of subsection (1) of section 34 of this Act to the tenant shall be construed as including the beneficiaries or beneficiary; and

(c) a change in the persons of the trustees shall not be treated as a change in the person of the tenant.

(2) Where the landlord's interest is held on trust the references in paragraph (g) of subsection (1) of section 30 of this Act to the landlord shall be construed as including references to the beneficiaries under the trust or any of them; but, except in the case of a trust arising under a will or on the intestacy of any person, the reference in subsection (2) of that section to the creation of the interest therein mentioned shall be construed as including the creation of the trust.

Partnerships

41A–(1) The following provisions of this section shall apply where-

(a) a tenancy is held jointly by two or more persons (in this section referred to as the joint tenants); and

(b) the property comprised in the tenancy is or includes premises occupied for the purposes of a business; and

(c) the business (or some other business) was at some time during the existence of the tenancy carried on in partnership by all the persons who were then the joint tenants or by those and other persons and the joint tenants' interest in the premises was then partnership property; and

(d) the business is carried on (whether alone or in partnership with other persons) by one or some only of the joint tenants and no part of the property comprised in the tenancy is occupied, in right of the tenancy, for the purposes of a business carried on (whether alone or in partnership with other persons) by the other or others.

(2) In the following provisions of this section those of the joint tenants who for the time being carry on the business are referred to as the business tenants and the others as the other joint tenants.

(3) Any notice given by the business tenants which, had it been given by all the joint tenants, would have been—

(a) a tenant's request for anew tenancy made in accordance with section 26 of this Act; or

(b) a notice under subsection (1) or subsection (2) of section 27 of this Act; shall be treated as such if it states that it is given by virtue of this section and sets out the facts by virtue of which the persons giving it are the business tenants;

and references in those sections and in section 24A of this Act to the tenant shall be construed accordingly.

(4) A notice given by the landlord to the business tenants which, had it been given to all the joint tenants, would have been a notice under section 25 of this Act shall be treated as such a notice, and references in that section to the tenant shall be construed accordingly.

(5) An application under section 24(1) of this Act for a new tenancy may, instead of being made by all the joint tenants, be made by the business tenants alone; and where it is so made-

(a) this Part of this Act shall have effect, in relation to it, as if the references therein to the tenant included references to the business tenants alone; and

(b) the business tenants shall be liable, to the exclusion of the other joint tenants, for the payment of rent and the discharge of any other obligation under the current tenancy for any rental period beginning after the date specified in the landlord's notice under section 25 of this Act or, as the case my be, beginning on or after the date specified in their request for a new tenancy.

(6) Where the court makes an order under ~~section 29(1) of this Act for the grant of a new tenancy on an application made by the business tenants it may order the grant to be made to them or to them jointly~~ **section 29 of this Act for the grant of a new tenancy it may order the grant to be made to the business tenants or to them jointly** with the persons carrying on the business in partnership with them, and may order the grant to be made subject to the satisfaction, within a time specified by the order, of such conditions as to guarantors, sureties or otherwise as appear to the court equitable, having regard to the omission of the other joint tenants from the persons who will be the tenants under the new tenancy.

(7) The business tenants shall be entitled to recover any amount payable by way of' compensation under section 37 or section 59 of this Act.

Groups of companies

42–(1) For the purposes of this section two bodies corporate shall be taken to be members of a group if and only if one is a subsidiary of the other or both are subsidiaries of the third body corporate **or the same person has a controlling interest in both**.

~~In this subsection "subsidiary" has the same meaning given by section 736 of the Companies Act 1985.~~

(2) Where a tenancy is held by a member of a group, occupation by another member of the group, and the carrying on of a business by another member of the group, shall be treated for the purposes of section 23 of this Act as equivalent to occupation or the carrying on of a business by the member of the group holding the tenancy; and in relation to a tenancy to which this Part of this Act applies by virtue of the foregoing provisions of this subsection—

(a) references (however expressed) in this Part of this Act and in the Ninth Schedule to this Act to the business of or to use occupation or enjoyment by the tenant shall be construed as including references to the business of or to use occupation or enjoyment by the said other member;

(b) the reference in paragraph (d) of subsection (1) of section 34 of this Act to the tenant shall be construed as including the said other member; and

(c) an assignment of the tenancy from one member of the group to another shall not be treated as a change in the person of the tenant.

(3) Where the landlord's interest is held by a member of a group—

(a) the reference in paragraph (g) of subsection (1) of section 30 of this Act to intended occupation by the landlord for the purposes of a business to be carried on by him shall be construed as including intended occupation by any member of the group for the purposes of a business to be carried on by that member; and

(b) the reference in subsection (2) of that section to the purchase or creation of any interest shall be construed as a reference to a purchaser from or creation by a person other than a member of the group.

Tenancies excluded from Part II

43–(1) This Part of this Act does not apply—

(a) to a tenancy of an agricultural holding which is a tenancy in relation to which the Agricultural Holdings Act 1986 applies or a tenancy which would be a tenancy of an agricultural holding in relation to which that Act applied if subsection (3) of section 2 of that Act did not have effect or, in a case where approval was given under subsection (1) of that section, if that approval had not been given;

(aa) to a farm business tenancy;

(b) to a tenancy created by a mining lease; or

(c) [Repealed]

(d) [Repealed]

(2) This Part of this Act does not apply to a tenancy granted by reason that the tenant was the holder of an office, appointment or employment from the grantor thereof and continuing only so long as the tenant holds the office, appointment or employment, or terminable by the grantor on the tenant's ceasing to hold it, or coming to an end at a time fixed by reference to the time at which the tenant ceases to hold it:

Provided that this subsection shall not have effect in relation to a tenancy granted after the commencement of this Act unless the tenancy was granted by an instrument in writing which expressed the purpose for which the tenancy was granted.

(3) This Part of this Act does not apply to a tenancy granted for a term certain not exceeding six months unless—

(a) the tenancy contains provision for renewing the term or for extending it beyond six months from its beginning; or

(b) the tenant has been in occupation for a period which, together with any period during which any predecessor in the carrying on of the business carried on by the tenant was in occupation, exceeds twelve months.

Jurisdiction of county court to make declaration

43A. Where the rateable value of the holding is such that the jurisdiction conferred on the court by any other provision of this Part of this Act is, by virtue of section 63 of this Act, exercisable by the county court, the county court shall have jurisdiction (but without prejudice to the jurisdiction of the High Court) to make any declaration as to any matter arising under this Part of this Act, whether or not any other relief is sought in the proceedings.

Meaning of "the landlord," in Part II, and provisions as to mesne landlords, etc.

44–(1) Subject to ~~the next following subsection~~ **subsections (1A) and (2) below**, in this Part of this Act the expression "the landlord" in relation to a tenancy (in this section referred to as "the relevant tenancy"), means the person (whether or not he is the immediate landlord) who is the owner of that interest in the property comprised in the relevant tenancy which for the time being fulfils the following conditions, that is to say—

(a) that it is an interest in reversion expectant (whether immediately or not) on the termination of the relevant tenancy, and

(b) that it is either the fee simple or a tenancy which will not come to an end within fourteen months by effluxion of time and, if it is such a tenancy, that no notice has been given by virtue of which it will come to an end within fourteen months or any further time by which it may be continued under section 36(2) or section 64 of this Act, and is not itself in reversion expectant (whether immediately or not) on an interest which fulfils those conditions.

(1A) The reference in subsection (1) above to a person who is the owner of an interest such as is mentioned in that subsection is to be construed, where different persons own such interests in different parts of the property, as a reference to all those persons collectively.

(2) References in this Part of this Act to a notice to quit given by the landlord are references to a notice to quit given by the immediate landlord.

(3) The provisions of the Sixth Schedule to this Act shall have effect for the application of this Part of this Act to cases where the immediate landlord of the tenant is not the owner of the fee simple in respect of the holding.

45 [Repealed]

Interpretation of Part II

46–(1) In this Part of this Act—

"business" has the meaning assigned to it by subsection (2) of section 23 of this Act;

~~"current tenancy" has the meaning assigned to it by subsection (1) of section 26 of this Act~~; **"current tenancy" means the tenancy under which the tenant holds for the time being;**

"date of termination" has the meaning assigned to it by subsection (1) of section 25 of this Act;

subject to the provisions of section 32 of this Act, "the holding" has the meaning assigned to it by subsection (3) of section 23 of this Act;

"interim rent" has the meaning given by section 24A(1) of this Act;

"mining lease" has the same meaning as in the Landlord and Tenant Act 1927.

(2) For the purposes of this Part of this Act, a person has a controlling interest in a company if, had he been a company, the other company would have been its subsidiary; and in this Part—

"company" has the meaning given by section 735 of the Companies Act 1985; and

"subsidiary" has the meaning given by section 736 of that Act.

PART III

COMPENSATION FOR IMPROVEMENTS

Time for making claims for compensation for improvements

47–(1) Where a tenancy is terminated by notice to quit, whether given by the landlord or by the tenant, or by a notice given by any person under Part I or Part II of this Act, the time for making a claim for compensation at the termination of the tenancy shall be a time falling within the period of three months beginning on the date on which the notice is given:

Provided that where the tenancy is terminated by a tenant's request for a new tenancy under section 26 of this Act, the said time shall be a time falling within the period of three months beginning on the date on which the landlord gives notice, or (if he has not given such a notice) the latest date on which he could have given notice, under subsection (6) of the said section 26 or, as the case may be, paragraph (a) of subsection (4) of section 57 or paragraph (b) of subsection (1) of section 58 of this Act.

(2) Where a tenancy comes to an end by effluxion of time, the time for making such a claim shall be a time not earlier than six nor later than three months before the coming to an end of the tenancy.

(3) Where a tenancy is terminated by forfeiture or re-entry, the time for making such a claim shall be a time falling within the period of three months beginning with the effective date of the order of the court for the recovery of possession of the land comprised in the tenancy or, if the tenancy is terminated by re-entry without such an order, the period of three months beginning with the date of the re-entry.

(4) In the last foregoing subsection the reference to the effective date of an order is a reference to the date on which the order is to take effect according to the terms thereof or the date on which it ceases to be subject to appeal, which ever is the later.

(5) In subsection (1) of section 1 of the Act of 1927, for paragraphs (a) and (b) (which specify the time for making claims for compensation) there shall be substituted the words "and within the time limited by section 47 of the Landlord and Tenant Act 1954."

Amendments as to limitations on tenant's right to compensation

48-(1) So much of paragraph (b) of subsection (1) of section 2 of the Act of 1927 as provides that a tenant shall not be entitled to compensation in respect of any improvement made in pursuance of a statutory obligation shall not apply to any improvement begun after the commencement of this Act, but section 3 of the Act of 1927 (which enables a landlord to object to a proposed improvement) shall not have effect in relation to an improvement made in pursuance of a statutory obligation except so much thereof as—

- (a) requires the tenant to serve on the landlord notice of his intention to make the improvement together with such a plan and specification as are mentioned in that section and to supply copies of the plan and specification at the request of any superior landlord; and
- (b) enables the tenant to obtain at his expense a certificate from the landlord or the tribunal that the improvement has been duly executed.

(2) Paragraph (c) of the said subsection (1) (which provides that a tenant shall not be entitled to compensation in respect of any improvement made less than three years before the termination of the tenancy) shall not apply to any improvement begun after the commencement of this Act.

(3) No notice shall be served after the commencement of this Act under paragraph (d) of the said subsection (1) (which excludes rights to compensation where the landlord serves on the tenant notice offering a renewal of the tenancy on reasonable terms).

Restrictions on contracting out

49 In section 9 of the Act of 1927 (which provides that Part I of that Act shall apply notwithstanding any contract to the contrary made after the date specified in that section) the proviso (which requires effect to be given to such a contract where it appears to the tribunal that the contract was made for adequate consideration) shall cease to have effect except as respects a contract made before the tenth day of December, nineteen hundred and fifty-three.

Interpretation of Part III

50 In this Part of this Act the expression "Act of 1927" means the Landlord and Tenant Act 1927, the expression "compensation" means compensation under Part I of that Act in respect of an improvement, and other expressions used in this Part of this Act and in the Act of 1927 have the same meanings in this Part of this Act as in that Act.

[Sections 51 to 54 not reproduced here]

PART IV

MISCELLANEOUS AND SUPPLEMENTARY

Compensation for possession obtained by misrepresentation

55 ~~(1) Where under Part I of this Act an order is made for possession of the property comprised in a tenancy, or under Part II of this Act the court refuses an order for the grant of a new tenancy, and it is subsequently made to appear to the court that the order was obtained, or the court induced to refuse the grant, by misrepresentation or the concealment of material facts, the court may order the landlord to pay to the tenant such sum as appears sufficient as compensation for damage or loss sustained by the tenant as the result of the order or refusal.~~

~~(2) In this section the expression "the landlord" means the person applying for possession or opposing an application for the grant of a new tenancy and the expression "the tenant" means the person against whom the order for possession was made or to whom the grant of a new tenancy was refused.~~

Application to Crown

56–(1) Subject to the provisions of this and the four next following sections, Part II of this Act shall apply where there is an interest belonging to Her Majesty in right of the Crown or the Duchy of Lancaster or belonging to the Duchy of Cornwall, or belonging to a Government department or held on behalf of Her Majesty for the purposes of a Government department, in like manner as if that interest were an interest not so belonging or held.

(2) The provisions of the Eighth Schedule to this Act shall have effect as respects the application of Part II of this Act to cases where the interest of the landlord belongs to Her Majesty in right of the Crown or the Duchy of Lancaster or to the Duchy of Cornwall.

(3) Where a tenancy is held by or on behalf of a Government department and the property comprised therein is or includes premises occupied for any purposes of a Government department, the tenancy shall be one to which Part II of this Act applies; and for the purposes of any provision of the said Part II or the Ninth Schedule to this Act which is applicable only if either or both of the following conditions are satisfied, that is to say—

(a) that any premises have during any period been occupied for the purposes of the tenant's business;

(b) that on any change of occupier of any premises the new occupier succeeded to the business of the former occupier,

the said conditions shall be deemed to be satisfied respectively, in relation to such a tenancy, if during that period or, as the case may be, immediately before and immediately after the change, the premises were occupied for the purposes of a Government department.

(4) The last foregoing subsection shall apply in relation to any premises provided by a Government department without any rent being payable to the department therefor as if the premises were occupied for the purposes of a Government department.

(5) The provisions of Parts III and IV of this Act, amending any other enactment which binds the Crown or applies to land belonging to Her Majesty in right of the Crown or the Duchy of Lancaster, or land belonging to the Duchy of Cornwall, or to land belonging to any Government department, shall bind the Crown or apply to such land.

(6) Sections 53 and 54 of this Act shall apply where the interest of the landlord, or any other interest in the land in question, belongs to Her Majesty in right of the Crown or the Duchy of Lancaster or to the Duchy of Cornwall, or belongs to a Government department or is held on behalf of Her Majesty for the purposes of a Government department, in like manner as if that interest were an interest not so belonging or held.

(7) Part I of this Act shall apply where—

(a) there is an interest belonging to Her Majesty in right of the Crown and that interest is under the management of the Crown Estate Commissioners;

or

(b) there is an interest belonging to Her Majesty in right of the Duchy of Lancaster or belonging to the Duchy of Cornwall;

as if it were an interest not so belonging.

Modification on grounds of public interest of rights under Part II

57–(1) Where the interest of the landlord or any superior landlord in the property comprised in any tenancy belongs to or is held for the purposes of a Government department or is held by a local authority, statutory undertakers or a development corporation, the Minister or Board in charge of any Government department may certify that it is requisite for the purposes of the first-mentioned department, or, as the case may be, of the authority, undertakers or corporation, that the use or occupation of the property or a part thereof shall be changed by a specified date.

(2) A certificate under the last foregoing subsection shall not be given unless the owner of the interest belonging or held as mentioned in the last foregoing subsection has given to the tenant a notice stating—

(a) that the question of the giving of such a certificate is under consideration by the Minister or Board specified in the notice, and

(b) that if within twenty-one days of the giving of the notice the tenant makes to that Minister or Board representations in writing with respect to that question, they will be considered before the question is determined, and if the tenant makes any such representations within the said twenty-one days the Minister or Board shall consider them before determining whether to give the certificate.

(3) Where a certificate has been given under subsection (1) of this section in relation to any tenancy, then—

(a) if a notice given under subsection (1) of section 25 of this Act specifies as the date of termination a date not earlier than the date specified in the certificate and contains a copy of the certificate ~~subsections (5) and~~ **subsection** (6) of that section shall not apply to the notice and no application for a new tenancy shall be made by the tenant under **subsection (1) of** section 24 of this Act;

(b) if such a notice specifies an earlier date as the date of termination and contains a copy of the certificate, then if the court makes an order under Part II of this Act for the grant of a new tenancy the new tenancy shall be for a term expiring not later than the date specified in the certificate and shall not be a tenancy to which Part II of this Act applies.

(4) Where a tenant makes a request for a new tenancy under section 26 of this Act, and the interest of the landlord or any superior landlord in the property comprised in the current tenancy belongs or is held as mentioned in subsection (1) of this section, the following provisions shall have effect:—

(a) if a certificate has been given under the said subsection (1) in relation to the current tenancy, and within two months after the making of the request the landlord gives notice to the tenant that the certificate has been given and the notice contains a copy of the certificate, then,—

(i) if the date specified in the certificate is not later than that specified in the tenant's request for a new tenancy, the tenant shall not make an application under section 24 of this Act for the grant of a new tenancy;

(ii) if, in any other case, the court makes an order under Part II of this Act for the grant of a new tenancy the new tenancy shall be for a term expiring not later than the date specified in the certificate and shall not be a tenant to which Part II of this Act applies;

(b) if no such certificate has been given but notice under subsection (2) of this section has been given before the making of the request or within two months thereafter, the request shall not have effect, without prejudice however, to the making of a new request when the Minister or Board has determined whether to give a certificate.

(5) Where application is made to the court under Part II of this Act for the grant of a new tenancy and the landlord's interest in the property comprised in the tenancy belongs or is held as mentioned in subsection (1) of this section, the Minister or Board in charge of any Government department may certify that it is necessary in the public interest that if the landlord makes an application in that behalf the court shall determine as a term of the new tenancy that is shall be terminable by six months' notice to quit given by the landlord.

Subsection (2) of this section shall apply in relation to a certificate under this subsection, and if notice under the said subsection (2) has been given to the tenant—

(a) the court shall not determine the application for the grant of a new tenancy until the Minister or Board has determined whether to give a certificate,

(b) if a certificate is given, the court shall on the application of the landlord determine as a term of the new tenancy that it shall be terminable as aforesaid, and section 25 of this Act shall apply accordingly.

(6) The foregoing provisions of this section shall apply to an interest held by a Health Authority or Special Health Authority as they apply to an interest held by a local authority but with the substitution, for the reference to the purposes of the authority, of a reference to the purposes of the National Health Service Act 1977.

(7) Where the interest of the landlord or any superior landlord in the property comprised in any tenancy belongs to the National Trust the Minister of Works may certify that it is requisite, for the purpose of securing that the property will as from a specified date be used or occupied in a manner better suited to the nature thereof, that the use or occupation of the property should be changed; and subsections (2) to (4) of this section shall apply in relation to certificates under this subsection, and to cases where the interest of the landlord or any superior landlord belongs to the National Trust, as those subsections apply in relation to certificates under subsection (1) of this section and to cases where the interest of the landlord or any superior landlord belongs or is held as mentioned in that subsection.

(8) In this and the next following section the expression "Government department" does not include the Commissioners of Crown Lands and the expression "landlord" has the same meaning as in Part II of this Act; and in the last foregoing subsection the expression "National Trust" means the National Trust for Places of Historic Interest or Natural Beauty.

Termination on special grounds of tenancies to which Part II applies

58–(1) Where the landlord's interest in the property comprised in any tenancy belongs or is held for the purposes of a Government department, and the Minister or Board in charge of any Government department certifies that for reasons of national security it is necessary that the use or occupation of the property should be discontinued or changed, then—

(a) if the landlord gives a notice under subsection (1) of section 25 of this Act containing a copy of the certificate, ~~subsections (5) and~~ **subsection** (6) of that section shall not apply to the notice and no application for a new tenancy shall be made by the tenant under **subsection (1) of** section 24 of this Act;

(b) if (whether before or after the giving of the certificate) the tenant makes a request for a new tenancy under section 26 of this Act, and within two months after the making the request the landlord gives notice to the tenant that the certificate has been given and the notice contains a copy of the certificate—

(i) the tenant shall not make an application under section 24 of this Act for the grant of a new tenancy, and

(ii) if the notice specifies as the date on which the tenancy is to terminate a date earlier than that specified in the tenant's request as the date on which the new tenancy is to begin but neither earlier than six months from the giving of the notice nor earlier than the earliest date at which apart from this Act the tenancy would come to an end or could be brought to an end, the tenancy shall terminate on the date specified in the notice instead of that specified in the request.

(2) Where the landlord's interest in the property comprised in any tenancy belongs to or is held for the purposes of a Government department, nothing in this Act shall invalidate an agreement to the effect—

(a) that on the giving of such a certificate as is mentioned in the last foregoing subsection the tenancy may be terminated by notice to quit given by the landlord of such length as may be specified in the agreement, if the notice contains a copy of the certificate; and

(b) that after the giving of such a notice containing such a copy the tenancy shall not be one to which Part II of this Act applies.

(3) Where the landlord's interest in the property comprised in any tenancy is held by statutory undertakers, nothing in this Act shall invalidate an agreement to the effect—

(a) that where the Minister or Board in charge of a Government department certifies that possession of the property comprised in the tenancy or a part thereof is urgently required for carrying out repairs (whether on that property or elsewhere) which are needed for the proper operation of the landlord's undertaking, the tenancy may be terminated by notice to quit given by the landlord of such length as may be specified in the agreement, if the notice contains a copy of the certificate; and

(b) that after the giving of such a notice containing such a copy, the tenancy shall not be one to which Part II of this Act applies.

(4) Where the court makes an order under Part II of this Act for the grant of a new tenancy and the Minister or Board in charge of any Government department certifies that the public interest requires the tenancy to be subject to such a term as is mentioned in paragraph (a) or (b) of this subsection, as the case may be, then—

(a) if the landlord's interest in the property comprised in the tenancy belongs to or is held for the purposes of a Government department, the court shall on the application of the landlord determine as a term of the new tenancy that such an agreement as is mentioned in subsection (2) of this section and specifying such length of notice as is mentioned in the certificate shall be embodied in the new tenancy;

(b) if the landlord's interest in that property is held by statutory undertakers, the court shall on the application of the landlord determine as a term of the new tenancy that such an agreement as is mentioned in subsection (3) of this section and specifying such length of notice as is mentioned in the certificate shall be embodied in the new tenancy.

Compensation for exercise of powers under sections 57 and 58

59–(1) Where by virtue of any certificate given for the purposes of either of the two last foregoing sections or, subject to subsection (1A) below, section 60A below the tenant is precluded from obtaining an order for the grant of a new tenancy, or of a new tenancy for a term expiring later than a specified date, the tenant shall be entitled on quitting the premises to recover from the owner of the interest by virtue of which the certificate was given an amount by way of compensation, and subsections (2), (3) **to (3B)** and (5) to (7) of section 37 of this Act shall with the necessary modifications apply for the purposes of ascertaining the amount.

(1A) No compensation shall be recoverable under subsection (1) above where the certificate was given under section 60A below and either—

(a) the premises vested in the Welsh Development Agency under section 7 (property of Welsh Industrial Estates Corporation) or 8 (land held under Local Employment Act 1972) of the Welsh Development Agency Act 1975, or

(b) the tenant was not tenant of the premises when the said Agency acquired the interest by virtue of which the certificate was given.

(2) Subsections (2) and (3) of section 38 of this Act shall apply to compensation under this section as they apply to compensation under section 37 of this Act.

Special provisions as to premises in development or intermediate areas

60–(1) Where the property comprised in a tenancy consists of premises of which the Secretary of State or the Urban Regeneration Agency is the landlord, being premises situated in a locality which is either—

(a) a development area; or

(b) an intermediate area;

and the Secretary of State certifies that it is necessary or expedient for achieving the purpose mentioned in section 2(1) of the Local Employment Act 1972 that the use or occupation of the property should be changed, paragraphs (a) and (b) of subsection (1) of section 58 of this Act shall apply as they apply where such a certificate is given as is mentioned in that subsection.

(2) Where the court makes an order under Part II of this Act for the grant of a new tenancy of any such premises as aforesaid, and the Secretary of State certifies that it is necessary or expedient as aforesaid that the tenancy should be subject to a term, specified in the certificate, prohibiting or restricting the tenant from assigning the tenancy or sub-letting, charging or parting with possession of the premises or any part thereof or changing the use of the premises or any part thereof, the court shall determine that the terms of the tenancy shall include the terms specified in the certificate.

(3) In this section "development area" and "intermediate area" mean an area for the time being specified as a development area or, as the case may be, as an intermediate area by an order made, or having effect as if made, under section 1 of the Industrial Development Act 1982.

Welsh Development Agency premises

60A–(1) Where property comprised in a tenancy consists of premises of which the Welsh Development Agency is the landlord, and the Secretary of State certifies that it is necessary or expedient, for the purpose of providing employment appropriate to the needs of the area in which the premises are situated, that the use or occupation of the property should be changed, paragraphs (a) and (b) of section 58(l) above shall apply as they apply where such a certificate is given as is mentioned in that sub-section.

(2) Where the court makes an order under Part II of this Act for the grant of a new tenancy of any such premises as aforesaid, and the Secretary of State certifies that it is necessary or expedient as aforesaid that the tenancy should be subject to a term, specified in the certificate, prohibiting or restricting the tenant from assigning the tenancy or subletting, charging or parting with possession of the premises or any part of the premises or changing the use of the premises or any part of the premises, the court shall determine that the terms of the tenancy shall include the terms specified in the certificate.

60B to 62 [Repealed]

Jurisdiction of court for purposes of Parts I and II and of Part I of Landlord and Tenant Act 1927

63–(1) Any jurisdiction conferred on the court by any provision of Part I of this Act shall be exercised by the county court.

(2) Any jurisdiction conferred on the court by any provision of Part II of this Act or conferred on the tribunal by Part I of the Landlord and Tenant Act 1927, shall, subject to the provisions of this section, be exercised, by the High Court or a county court.

(3) [Repealed]

(4) The following provisions shall have effect as respects transfer of proceedings from or to the High Court or the county court, that is to say—

(a) where an application is made to the one but by virtue of an Order under section 1 of the Courts and Legal Services Act 1990, cannot be entertained except by the other, the application shall not be treated as improperly made but any proceedings thereon shall be transferred to the other court;

(b) any proceedings under the provisions of Part II of this Act or of Part I of the Landlord and Tenant Act 1927, which are pending before one of those courts may by order of that court made on the application of any person interested be transferred to the other court, if it appears to the court making the order that it is desirable that the proceedings and any proceedings before the other court should both be entertained by the other court.

(5) In any proceedings where in accordance with the foregoing provisions of this section the county court exercises jurisdiction the powers of the judge of summoning one or more assessors under subsection (1) of section 63 (1) of the County Courts Act 1984, may be exercised notwithstanding that no application is made in that behalf by any party to the proceedings.

(6) Where in any such proceedings an assessor is summoned by a judge under the said subsection (1),—

(a) he may, if so directed by the judge, inspect the land to which the proceedings relate without the judge and report to the judge in writing thereon;

(b) the judge may on consideration of the report and any observations of the parties thereon give such judgment or make such order in the proceedings as may be just;

(c) the remuneration of the assessor shall beat such rate as maybe determined by the Lord Chancellor with the approval of the Treasury and shall be defrayed out of moneys provided by Parliament.

(7) In this section the expression "the holding"—

a) in relation to proceedings under Part II of this Act, has the meaning assigned to it by subsection (3) of section 23 of this Act,

(b) in relation to proceedings under Part I of the Landlord and Tenant Act 1927, has the same meaning as in the said Part I.

(9) Nothing in this section shall prejudice the operation of section 41 of the County Courts Act 1984 (which relates to the removal into the High Court of proceedings commenced in a county court).

(10) In accordance with the foregoing provisions of this section, for section 21 of the Landlord and Tenant Act 1927, there shall be substituted the following section—

"The tribunal

21. The tribunal for the purposes of Part I of this Act shall be the court exercising jurisdiction in accordance with the provisions of section 63 of the Landlord and Tenant Act 1954".

64–(1) In any case where—

- (a) a notice to terminate a tenancy has been given under Part I or Part II of this Act or a request for a new tenancy has been made under Part II thereof, and
- (b) an application to the court has been made under the said Part I ~~or the said Part II~~, **under section 24(1) or 29(2) of this Act** as the case may be, and
- (c) apart from this section the effect of the notice or request would be to terminate the tenancy before the expiration of the period of three months beginning with the date on which the application is finally disposed of,

the effect of the notice or request shall be to terminate the tenancy at the expiration of the said period of three months and not at any other time.

(2) The reference in paragraph (c) of subsection (1) of this section to the date on which an application is finally disposed of shall be construed as a reference to the earliest date by which the proceedings on the application (including any proceedings on or in consequence of an appeal) have been determined and any time for appealing or further appealing has expired, except that if the application is withdrawn or any appeal is abandoned the reference shall be construed as a reference to the date of the withdrawal or abandonment.

Provisions as to reversions

65–(1) Where by virtue of any provision of this Act a tenancy (in this sub-section referred to as "the inferior tenancy") is continued for a period such as to extend to or beyond the end of the term of a superior tenancy, the superior tenancy shall, for the purposes of this Act and of any other enactment and of any rule of law, be deemed so long as it subsists to be an interest in reversion expectant upon the termination of the inferior tenancy and, if there is no intermediate tenancy, to be the interest in reversion immediately expectant upon the termination thereof.

(2) In the case of a tenancy continuing by virtue of any provision of this Act after the coming to an end of the interest in reversion immediately expectant upon the termination thereof, subsection (1) of section 139 of the Law of Property Act 1925 (which relates to the effect of the extinguishment of a reversion) shall apply as if references in the said subsection (1) to the surrender or merger of the reversion included references to the coming to an end of the reversion for any reason other than surrender or merger.

(3) Where by virtue of any provision of this Act a tenancy (in this subsection referred to as "the continuing tenancy") is continued beyond the beginning of a reversionary tenancy which was granted (whether before or after the commencement of this Act) so as to begin on or after the date on which apart from this Act the continuing tenancy would have come to an end, the reversionary tenancy shall have effect as if it had been granted subject to the continuing tenancy.

(4) Where by virtue of any provision of this Act a tenancy (in this subsection referred to as "the new tenancy") is granted for a period beginning on the same date as a reversionary tenancy or for a period such as to extend beyond the beginning of the term of a reversionary tenancy, whether the reversionary tenancy in question was granted before or after the commencement of this Act, the reversionary tenancy shall have effect as if it had been granted subject to the new tenancy.

Provisions as to notices

66–(1) Any form of notice required by this Act to be prescribed shall be prescribed by regulations made by the Secretary of State by statutory instrument.

(2) Where the form of a notice to be served on persons of any description is to be prescribed for any of the purposes of this Act, the form to be prescribed shall include such an explanation of the relevant provisions of this Act as appears to the Secretary of State requisite for informing persons of that description of their rights and obligations under those provisions.

(3) Different forms of notice may be prescribed for the purposes of the operation of any provision of this Act in relation to different cases.

(4) Section 23 of the Landlord and Tenant Act 1927 (which relates to the service of notices) shall apply for the purposes of this Act.

(5) Any statutory instrument under this section shall be subject to annulment in pursuance of a resolution of either House of Parliament.

Provisions as to mortgagees in possession

67 Anything authorised or required by the provisions of this Act, other than subsection (2) or (3) of section 40, to be done at any time by, to or with the landlord, or a landlord of a specified description, shall, if at that time the interest of the landlord in question is subject to a mortgage and the mortgagee is in possession or a receiver appointed by the mortgagee or by the courts is in receipt of the rents and profits, be deemed to be authorised or required to be done by, to or with the mortgagee instead of that landlord.

68 [Not reproduced here]

Interpretation

69–(1) In this Act, the following expressions have the meanings hereby assigned to them respectively, that is to say—

"agricultural holding" has the same meaning as in the Agricultural Holdings Act 1986;

"development corporation" has the same meaning as in the New Towns Act 1946;

"farm business tenancy" has the same meaning as in the Agricultural Tenancies Act 1995;

"local authority" means any local authority within the meaning of the Town and Country Planning Act 1990, any National Park Authority, the Broads Authority or joint authority established by Part 4 of the Local Government Act 1985;

"mortgage" includes a charge or lien and "mortgagor" and "mortgagee" shall be construed accordingly;

"notice to quit" means a notice to terminate a tenancy (whether a periodical tenancy or a tenancy for a term of years certain) given in accordance with the provisions (whether express or implied) of that tenancy;

"repairs" includes any work of maintenance, decoration or restoration, and references to repairing, to keeping or yielding up in repair and to state of repair shall be construed accordingly;

"statutory undertakers" has the same meaning as in the Town and Country Planning Act 1990;

"tenancy" means a tenancy created either immediately or derivatively out of the freehold, whether by a lease or underlease, by an agreement for a lease or underlease or by a tenancy agreement or in pursuance of any enactment (including this Act), but does not include a mortgage term or any interest arising in favour of a mortgagor by his attorning tenant to his mortgagee, and references to the granting of a tenancy and to demised property shall be construed accordingly;

"terms", in relation to a tenancy, includes conditions.

(2) References in this Act to an agreement between the landlord and the tenant (except in section 17 and subsections (1) and (2) of section 38 thereof) shall be construed as references to an agreement in writing between them.

(3) Reference in this Act to an action for any relief shall be construed as including references to a claim for that relief by way of counterclaim in any proceedings.

Short title and citation, commencement and extent

70–(1) This Act may be cited as the Landlord and Tenant Act, 1954, and the Landlord and Tenant Act, 1927, and this Act may be cited together as the Landlord and Tenant Acts, 1927 and 1954.

(2) This Act shall come into operation on the first day of October, nineteen hundred and fifty-four.

(3) This Act shall not extend to Scotland or to Northern Ireland.

SCHEDULE 6

Section 44

PROVISIONS FOR PURPOSES OF PART II WHERE IMMEDIATE LANDLORD IS NOT THE FREEHOLDER

Definitions

1. In this Schedule the following expressions have the meanings hereby assigned to them in relation to a tenancy (in this Schedule referred to as "the relevant tenancy"), that is to say:

"the competent landlord" means the person who in relation to the tenancy is for the time being the landlord (m defined by section forty-four of this Act) for the purposes of Part II of this Act;

"mesne landlord" means a tenant whose interest is intermediate between the relevant tenancy and the interest of the competent landlord; and

"superior landlord" means a person (whether the owner of the fee simple or a tenant) whose interest is superior to the interest of the competent landlord.

Power of court to order reversionary tenancies

Where the period for which in accordance with the provisions of Part II of this Act it is agreed or determined by the court that a new tenancy should be granted thereunder will emend beyond the date on which the interest of the immediate landlord will come to an end, the power of the court under Part II of this Ad to order such a grant shall include power to order the grant of a new tenancy until the expiration of that interest and also to order the grant of such a reversionary tenancy or reversionary tenancies as may be required to secure that the combined effects of those grants will be equivalent to the grant of a tenancy for that period; and the provisions of Part II of this Act shall, subject to the necessary modifications, apply in relation to the grant of a tenancy together with one or more reversionary tenancies as they apply in relation to the grant of one new tenancy.

Acts of competent landlord binding on other landlords

3.(1) Any notice given by the competent landlord under Part II of this Act to terminate the relevant tenancy, and any agreement made between that landlord and the tenant as to the granting, duration, or terms of a future tenancy, being an agreement made for the purposes of the said Part II, shall bind the interest of any mesne landlord notwithstanding that be has not consented to the giving of the notice or was not a party to the agreement.

(2) The competent landlord shall have power for the purposes of Part II of this Act to give effect to any agreement with the tenant for the grant of a new tenancy beginning with the coming to an end of the relevant tenancy notwithstanding that the competent landlord will not be the immediate landlord at the commencement of the new tenancy, and any instrument made in the exercise of the power conferred by this sub-paragraph shall have effect as if the mesne landlord had been a party thereto.

(3) Nothing in the foregoing provisions of this paragraph shall prejudice the provisions of the next following paragraph.

Provisions as to consent of mesne landlord to acts of competent landlord

4.(1) If the competent landlord, not being the immediate landlord, gives any such notice or makes any such agreement as is mentioned in sub-paragraph (1) of the last foregoing paragraph without the consent of every mesne landlord, any mesne landlord whose consent has not been given thereto shall be entitled to compensation from the competent landlord for any loss arising in consequence of the giving of the notice or the making of the agreement.

(2) If the competent landlord applies to any mesne landlord for his consent to such a notice or agreement, that consent shall not be unreasonably withheld, but may be given subject to any conditions which may be reasonable (including conditions as to the modification of the proposed notice or agreement or as to the payment of compensation by the competent landlord).

(3) Any question arising under this paragraph whether consent has been unreasonably withheld or whether any conditions imposed on the giving of consent are unreasonable shall be determined by the court.

Consent of superior landlord required for agreements affecting his interest

5. An agreement between the competent landlord and the tenant made for the purposes of Part II of this Act in a case where:

(a) the competent landlord is himself a tenant; and

(b) the agreement would apart from this paragraph operate as respects any period after the coming to an end of the interest of the competent landlord;

shall not have effect unless every superior landlord who will be the immediate landlord of the tenant during any part of that period is a party to the agreement.

Withdrawal by competent landlord of notice given by mesne landlord

6. Where the competent landlord has given a notice under section 25 of this Act to terminate the relevant tenancy and, within two months after the giving of the notice, a superior landlord:

(a) becomes the competent landlord; and

(b) gives to the tenant notice in the prescribed form that he withdraws the notice, previously given;

the notice under section 25 of this Act shall cease to have effect, but without prejudice to the giving of a further notice under that section by the competent landlord.

Duty to inform superior landlords

7. If the competent landlord's interest in the property comprised in the relevant tenancy is a tenancy which will come or can be brought to an end within sixteen months (or any further time by which it may be continued under section 36(2) or section 64 of this Act) and he gives to the tenant under the relevant tenancy a notice under section 25 of this Act to terminate the tenancy or is given by him a notice under section 26(3) of this Act:

(a) the competent landlord shall forthwith send a copy of the notice to his immediate landlord; and

(b) any superior landlord whose interest in the property is a tenancy shall forthwith send to his immediate landlord any copy which has been sent to him in pursuance of the preceding sub-paragraph or this sub-paragraph.

SCHEDULE 8

Section 56

APPLICATION OF PART II TO LAND BELONGING TO CROWN AND DUCHIES OF LANCASTER AND CORNWALL

1. Where an interest in any property comprised in a tenancy belongs to Her Majesty in right of the Duchy of Lancaster, then for the purposes of Part II of this Act the Chancellor of the Duchy shall represent Her Majesty and shall be deemed to be the owner of the interest.

2. Where an interest in any property comprised in a tenancy belongs to the Duchy of Cornwall then for the purposes of Part II of this Act such person as the Duke of Cornwall, or other the possessor for the time being of the Duchy of Cornwall, appoints shall represent the Duke of Cornwall or other the possessor aforesaid, and shall be deemed to be the owner of the interest and may do any act or thing under the said Part II which the owner of that interest is authorised or required to do thereunder.

3. ...

4. The amount of any compensation payable under section thirty-seven of this Act by the Chancellor of the Duchy of Lancaster shall be raised and paid as an expense incurred in improvement of land belonging to Her Majesty in right of the Duchy within section twenty-five of the Act of the fifty-seventh year of King George the Third, Chapter ninety-seven.

5. Any compensation payable under section thirty-seven of this Act by the person representing the Duke of Cornwall or other the possessor for the time being of the Duchy of Cornwall shall be paid, and advances therefor made, in the manner and subject to the provisions of section eight of the Duchy of Cornwall Management Act 1863 with respect to improvements of land mentioned in that section.

SCHEDULE 9

Sections 41, 42, 56, 68

TRANSITIONAL PROVISIONS

1, 2. ...

3. Where immediately before the commencement of this Act a person was protected by section seven of the Leasehold Property (Temporary Provisions) Act 1951, against the making of an order or giving of a judgment for possession or ejectment, the Rent Acts shall apply in relation to the dwelling-house to which that person's protection extended immediately before the commencement of this Act as if section fifteen of this Act had always had effect.

4. For the purposes of section twenty-six and subsection (2) of section forty of this Act a tenancy which is not such a tenancy as is mentioned in subsection (1) of the said section twenty-six but is a tenancy to which Part 11 of this Act applies and in respect of which the following conditions are satisfied, that is to say:

(a) that it took effect before the commencement of this Act at the coming to an end by effluxion of time or notice to quit of a tenancy which is such a tenancy as is mentioned in subsection (1) of the said section twenty-six or is by virtue of this paragraph deemed to be such a tenancy; and

(b) that if this Act had then been in force the tenancy at the coming to an end of which it took effect would have been one to which Part n of this Act applies; and

(c) that the tenant is either the tenant under the tenancy at the coming to an end of which it took effect or a successor to his business,

shall be deemed to be such a tenancy as is mentioned in subsection (1) of the said section twenty-six.

5.(1) A tenant under a tenancy which was-current at the commencement of this Act shall not in any case be entitled to compensation under section thirty-seven or fifty-nine of this Act unless at the date on which he is to quit the holding the holding or part thereof has continuously been occupied for the purposes of the carrying on of the tenant's business (whether by him or by any other person) for at least five years.

(2) Where a tenant under a tenancy which was current at the commencement of this Act would but for this sub-paragraph be entitled both to

(a) compensation under section thirty-seven or section fifty-nine of this Act; and

(b) compensation payable, under the provisions creating the tenancy, on the termination of the tenancy,

he shall be entitled, at his option, to the one or the other, but not to both.

6.(1) Where the landlord's interest in the property comprised in a tenancy which immediately before the commencement of this Act, was terminable by less than six months' notice to quit given by the landlord belongs to or is held for the purposes of a Government department or is held by statutory undertakers, the tenancy shall have effect as if that shorter length of notice were specified in such an agreement as is mentioned in subsection (2) or (3) of section fifty-eight of this Act, as the case may be, and the agreement were embodied in the tenancy. (2) The last foregoing sub-paragraph shall apply in relation to a tenancy where the landlord's interest belongs or is held as aforesaid and which, immediately before the commencement of this Act, was terminable by the landlord without notice as if the tenancy had then been terminable by one month's notice to quit given by the landlord. 8. Where at the commencement of this Act any proceedings are pending on an application made before the commencement of this Act to the tribunal under section five of the Landlord and Tenant Act 1927, no further step shall be taken in the proceedings except for the purposes of an order as to costs; and where the tribunal has made an interim order in the proceedings under subsection (13) of section five of that Act authorising the tenant to remain in possession of the property comprised in his tenancy for any period, the tenancy shall be deemed not to have come to an end before the expiration of that period, and section twenty-four of this Act shall have effect in relation to it accordingly.

9, 10 ...

11. Notwithstanding the repeal of Part II of the Leasehold Property (Temporary Provisions) Act 1951, where immediately before the commencement of this Act a tenancy was being continued by subsection (3) of section eleven of that Act it shall not come to end at the commencement of this Act, and section twenty-four of this Act shall have effect in relation to it accordingly.

Appendix 2

Schedules 1–4 of The Regulatory Reform (Business Tenancies) (England and Wales) Order 2003 SI 2003 No 3096

SCHEDULE 1

Article 22(2)

FORM OF NOTICE THAT SECTIONS 24 TO 28 OF THE LANDLORD AND TENANT ACT 1954 ARE NOT TO APPLY TO A BUSINESS TENANCY

To:

____

____

____

[*Name and address of tenant*]

From:

____

____

____

[*Name and address of landlord*]

[a]

IMPORTANT NOTICE

You are being offered a lease without security of tenure. Do not commit yourself to the lease unless you have read this message carefully and have discussed it with a professional adviser.

Business tenants normally have security of tenure — the right to stay in their business premises when the lease ends.

If you commit yourself to the lease you will be giving up these important legal rights.

- You will have **no right** to stay in the premises when the lease ends.
- Unless the landlord chooses to offer you another lease, you will need to leave the premises.
- You will be unable to claim compensation for the loss of your business premises, unless the lease specifically gives you this right.
- If the landlord offers you another lease, you will have no right to ask the court to fix the rent.

It is therefore important to get professional advice — from a qualified surveyor, lawyer or accountant — before agreeing to give up these rights.

If you want to ensure that you can stay in the same business premises when the lease ends, you should consult your adviser about another form of lease that does not exclude the protection of the Landlord and Tenant Act 1954.

If you receive this notice at least 14 days before committing yourself to the lease, you will need to sign a simple declaration that you have received this notice and have accepted its consequences, before signing the lease.

But if you do not receive at least 14 days notice, you will need to sign a "statutory" declaration. To do so, you will need to visit an independent solicitor (or someone else empowered to administer oaths).

Unless there is a special reason for committing yourself to the lease sooner, you may want to ask the landlord to let you have at least 14 days to consider whether you wish to give up your statutory rights. If you then decided to go ahead with the agreement to exclude the protection of the Landlord and Tenant Act 1954, you would only need to make a simple declaration, and so you would not need to make a separate visit to an independent solicitor.

SCHEDULE 2

Article 22(2)

REQUIREMENTS FOR A VALID AGREEMENT THAT SECTIONS 24 TO 28 OF THE LANDLORD AND TENANT ACT 1954 ARE NOT TO APPLY TO A BUSINESS TENANCY

1. The following are the requirements referred to in section 38A(3)(b) of the Act.

2. Subject to paragraph 4, the notice referred to in section 38A(3)(a) of the Act must be served on the tenant not less than 14 days before the tenant enters into the tenancy to which it applies, or (if earlier) becomes contractually bound to do so.

3. If the requirement in paragraph 2 is met, the tenant, or a person duly authorised by him to do so, must, before the tenant enters into the tenancy to which the notice applies, or (if earlier) becomes contractually bound to do so, make a declaration in the form, or substantially in the form, set out in paragraph 7.

4. If the requirement in paragraph 2 is not met, the notice referred to in section 38A(3)(a) of the Act must be served on the tenant before the tenant enters into the tenancy to which it applies, or (if earlier) becomes contractually bound to do so, and the tenant, or a person duly authorised by him to do so, must before that time make a statutory declaration in the form, or substantially in the form, set out in paragraph 8.

5. A reference to the notice and, where paragraph 3 applies, the declaration or, where paragraph 4 applies, the statutory declaration must be contained in or endorsed on the instrument creating the tenancy.

6. The agreement under section 38A(1) of the Act, or a reference to the agreement, must be contained in or endorsed upon the instrument creating the tenancy.

7. The form of declaration referred to in paragraph 3 is as follows—

I

(*name of declarant*) of

(*address*) declare that—

1. I/

(*name of tenant*) propose(s) to enter into a tenancy of premises at

---

(*address of premises*) for a term commencing on

---

2. I/The tenant propose(s) to enter into an agreement with

(*name of landlord*) that the provisions of sections 24 to 28 of the Landlord and Tenant Act 1954 (security of tenure) shall be excluded in relation to the tenancy.

3. The landlord has, not less than 14 days before I/the tenant enter(s) into the tenancy, or (if earlier) become(s) contractually bound to do so served on me/the tenant a notice in the form, or substantially in the form, set out in Schedule 1 to the Regulatory Reform (Business Tenancies) (England and Wales) Order 2003. The form of notice set out in that Schedule is reproduced below.

4. I have/The tenant has read the notice referred to in paragraph 3 above and accept(s) the consequences of entering into the agreement referred to in paragraph 2 above.

5. (*as appropriate*) I am duly authorised by the tenant to make this declaration.

DECLARED this

---

day of

To:

[*Name and address of tenant*]

From:

___

___

___

[*name and address of landlord*]

IMPORTANT NOTICE

You are being offered a lease without security of tenure. Do not commit yourself to the lease unless you have read this message carefully and have discussed it with a professional adviser.

Business tenants normally have security of tenure — the right to stay in their business premises when the lease ends.

If you commit yourself to the lease you will be giving up these important legal rights.

- You will have **no right** to stay in the premises when the lease ends.
- Unless the landlord chooses to offer you another lease, you will need to leave the premises.
- You will be unable to claim compensation for the loss of your business premises, unless the lease specifically gives you this right.
- If the landlord offers you another lease, you will have no right to ask the court to fix the rent.

It is therefore important to get professional advice — from a qualified surveyor, lawyer or accountant — before agreeing to give up these rights.

If you want to ensure that you can stay in the same business premises when the lease ends, you should consult your adviser about another form of lease that does not exclude the protection of the Landlord and Tenant Act 1954.

If you receive this notice at least 14 days before committing yourself to the lease, you will need to sign a simple declaration that you have received this notice and have accepted its consequences, before signing the lease.

But if you do not receive at least 14 days notice, you will need to sign a "statutory" declaration. To do so, you will need to visit an independent solicitor (or someone else empowered to administer oaths).

Unless there is a special reason for committing yourself to the lease sooner, you may want to ask the landlord to let you have at least 14 days to consider whether you wish to give up your statutory rights. If you then decided to go ahead with the agreement to exclude the protection of the Landlord and Tenant Act 1954, you would only need to make a simple declaration, and so you would not need to make a separate visit to an independent solicitor.

8. The form of statutory declaration referred to in paragraph 4 is as follows—

I

(*name of declarant*) of

(*address*) do solemnly and sincerely declare that—

1. I

(*name of tenant*) propose(s) to enter into a tenancy of premises at

(*address of premises*) for a term commencing on

2. I/The tenant propose(s) to enter into an agreement with (name of landlord) that the provisions of sections 24 to 28 of the Landlord and Tenant Act 1954 (security of tenure) shall be excluded in relation to the tenancy.

3. The landlord has served on me/the tenant a notice in the form, or substantially in the form, set out in Schedule 1 to the Regulatory Reform (Business Tenancies) (England and Wales) Order 2003. The form of notice set out in that Schedule is reproduced below.

4. I have/The tenant has read the notice referred to in paragraph 3 above and accept(s) the consequences of entering into the agreement referred to in paragraph 2 above.

5. (*as appropriate*) I am duly authorised by the tenant to make this declaration.

To:

[*Name and address of tenant*]

From:

[*name and address of landlord*]

IMPORTANT NOTICE

You are being offered a lease without security of tenure. Do not commit yourself to the lease unless you have read this message carefully and have discussed it with a professional adviser.

Business tenants normally have security of tenure — the right to stay in their business premises when the lease ends.

If you commit yourself to the lease you will be giving up these important legal rights.

- You will have **no right** to stay in the premises when the lease ends.
- Unless the landlord chooses to offer you another lease, you will need to leave the premises.
- You will be unable to claim compensation for the loss of your business premises, unless the lease specifically gives you this right.
- If the landlord offers you another lease, you will have no right to ask the court to fix the rent.

It is therefore important to get professional advice — from a qualified surveyor, lawyer or accountant — before agreeing to give up these rights.

If you want to ensure that you can stay in the same business premises when the lease ends, you should consult your adviser about another form of lease that does not exclude the protection of the Landlord and Tenant Act 1954.

If you receive this notice at least 14 days before committing yourself to the lease, you will need to sign a simple declaration that you have received this notice and have accepted its consequences, before signing the lease.

But if you do not receive at least 14 days notice, you will need to sign a "statutory" declaration. To do so, you will need to visit an independent solicitor (or someone else empowered to administer oaths).

Unless there is a special reason for committing yourself to the lease sooner, you may want to ask the landlord to let you have at least 14 days to consider whether you wish to give up your statutory rights. If you then decided to go ahead with the agreement to exclude the protection of the Landlord and Tenant Act 1954, you would only need to make a simple declaration, and so you would not need to make a separate visit to an independent solicitor.

AND I make this solemn declaration conscientiously believing the same to be true and by virtue of the Statutory Declaration Act 1835.

DECLARED at

____

this

____

day of

____

Before me

(*signature of person before whom declaration is made*)

A commissioner for oaths or A solicitor empowered to administer oaths or (*as appropriate*)

SCHEDULE 3

Article 22(2)

FORM OF NOTICE THAT AN AGREEMENT TO SURRENDER A BUSINESS TENANCY IS TO BE MADE

To: ______________________________

____

____

[*Name and address of tenant*]

From: ______________________________

____

____

[*name and address of landlord*]

IMPORTANT NOTICE FOR TENANT

Do not commit yourself to any agreement to surrender your lease unless you have read this message carefully and discussed it with a professional adviser.

Normally, you have the right to renew your lease when it expires. By committing yourself to an agreement to surrender, **you will be giving up this important statutory right.**

- You will **not** be able to continue occupying the premises beyond the date provided for under the agreement for surrender, **unless** the landlord chooses to offer you a further term (in which case you would lose the right to ask the court to determine the new rent). You will need to leave the premises.
- You will be unable to claim compensation for the loss of your premises, unless the lease or agreement for surrender gives you this right.

A qualified surveyor, lawyer or accountant would be able to offer you professional advice on your options.

You do not have to commit yourself to the agreement to surrender your lease unless you want to.

If you receive this notice at least 14 days before committing yourself to the agreement to surrender, you will need to sign a simple declaration that you have received this notice and have accepted its consequences, before signing the agreement to surrender.

But if you do not receive at least 14 days notice, you will need to sign a "statutory" declaration. To do so, you will need to visit an independent solicitor (or someone else empowered to administer oaths).

Unless there is a special reason for committing yourself to the agreement to surrender sooner, you may want to ask the landlord to let you have at least 14 days to consider whether you wish to give up your statutory rights. If you then decided to go ahead with the agreement to end your lease, you would only need to make a simple declaration, and so you would not need to make a separate visit to an independent solicitor.

SCHEDULE 4

Article 22(2)

REQUIREMENTS FOR A VALID AGREEMENT TO SURRENDER A BUSINESS TENANCY

1. The following are the requirements referred to in section 38A(4)(b) of the Act.

2. Subject to paragraph 4, the notice referred to in section 38A(4)(a) of the Act must be served on the tenant not less than 14 days before the tenant enters into the agreement under section 38A(2) of the Act, or (if earlier) becomes contractually bound to do so.

3. If the requirement in paragraph 2 is met, the tenant or a person duly authorised by him to do so, must, before the tenant enters into the agreement under section 38A(2) of the Act, or (if earlier) becomes contractually bound to do so, make a declaration in the form, or substantially in the form, set out in paragraph 6.

4. If the requirement in paragraph 2 is not met, the notice referred to in section 38A(4)(a) of the Act must be served on the tenant before the tenant enters into the agreement under section 38A(2) of the Act, or (if earlier) becomes contractually bound to do so, and the tenant, or a person duly authorised by him to do so, must before that time make a statutory declaration in the form, or substantially in the form, set out in paragraph 7.

5. A reference to the notice and, where paragraph 3 applies, the declaration or, where paragraph 4 applies, the statutory declaration must be contained in or endorsed on the instrument creating the agreement under section 38A(2).

6. The form of declaration referred to in paragraph 3 is as follows—

I

___

(*name of declarant*) of

___

(*address*) declare that—

1. I have/

___

(*name of tenant*) has a tenancy of premises at

___

(*address of premises*) for a term commencing on

___

2. I/The tenant propose(s) to enter into an agreement with

___

(*name of landlord*) to surrender the tenancy on a date or in circumstances specified in the agreement.

3. The landlord has not less than 14 days before I/the tenant enter(s) into the agreement referred to in paragraph 2 above, or (if earlier) become(s) contractually bound to do so, served on me/the tenant a notice in the form, or substantially in the form, set out in Schedule 3 to Regulatory Reform (Business Tenancies) (England and Wales) Order 2003. The form of notice set out in that Schedule is reproduced below.

4. I have/The tenant has read the notice referred to in paragraph 3 above and accept(s) the consequences of entering into the agreement referred to in paragraph 2 above.

5. (as appropriate) I am duly authorised by the tenant to make this declaration.

DECLARED this

___

day of

___

To:

___

___

___

[*Name and address of tenant*]

From:

____

____

____

[*name and address of landlord*]

IMPORTANT NOTICE FOR TENANT

Do not commit yourself to any agreement to surrender your lease unless you have read this message carefully and discussed it with a professional adviser.

Normally, you have the right to renew your lease when it expires. By committing yourself to an agreement to surrender, **you will be giving up this important statutory right.**

- You will **not** be able to continue occupying the premises beyond the date provided for under the agreement for surrender, **unless** the landlord chooses to offer you a further term (in which case you would lose the right to ask the court to determine the new rent). You will need to leave the premises.
- You will be unable to claim compensation for the loss of your premises, unless the lease or agreement for surrender gives you this right.

A qualified surveyor, lawyer or accountant would be able to offer you professional advice on your options.

You do not have to commit yourself to the agreement to surrender your lease unless you want to.

If you receive this notice at least 14 days before committing yourself to the agreement to surrender, you will need to sign a simple declaration that you have received this notice and have accepted its consequences, before signing the agreement to surrender.

But if you do not receive at least 14 days notice, you will need to sign a "statutory" declaration. To do so, you will need to visit an independent solicitor (or someone else empowered to administer oaths).

Unless there is a special reason for committing yourself to the agreement to surrender sooner, you may want to ask the landlord to let you have at least 14 days to consider whether you wish to give up your statutory rights. If you then decided to go ahead with the agreement to end your lease, you would only need to make a simple declaration, and so you would not need to make a separate visit to an independent solicitor.

7. The form of statutory declaration referred to in paragraph 4 is as follows—

I

____

(*name of declarant*) of

____

(*address*) do solemnly and sincerely declare that—

1. I have/

____

(*name of tenant*) has a tenancy of premises at

____

(*address of premises*) for a term commencing on

____

2. I/The tenant propose(s) to enter into an agreement with

____

(*name of landlord*) to surrender the tenancy on a date or in circumstances specified in the agreement.

3. The landlord has served on me/the tenant a notice in the form, or substantially in the form, set out in Schedule 3 to the Regulatory Reform (Business Tenancies) (England and Wales) Order 2003. The form of notice set out in that Schedule is reproduced below.

4. I have/The tenant has read the notice referred to in paragraph 3 above and accept(s) the consequences of entering into the agreement referred to in paragraph 2 above.

5. (as appropriate) I am duly authorised by the tenant to make this declaration.

To:

____

____

____

[*Name and address of tenant*]

From:

____

____

____

[*name and address of landlord*]

IMPORTANT NOTICE FOR TENANT

Do not commit yourself to any agreement to surrender your lease unless you have read this message carefully and discussed it with a professional adviser.

Normally, you have the right to renew your lease when it expires. By committing yourself to an agreement to surrender, **you will be giving up this important statutory right.**

- You will **not** be able to continue occupying the premises beyond the date provided for under the agreement for surrender, **unless** the landlord chooses to offer you a further term (in which case you would lose the right to ask the court to determine the new rent). You will need to leave the premises.
- You will be unable to claim compensation for the loss of your premises, unless the lease or agreement for surrender gives you this right.

A qualified surveyor, lawyer or accountant would be able to offer you professional advice on your options.

You do not have to commit yourself to the agreement to surrender your lease unless you want to.

If you receive this notice at least 14 days before committing yourself to the agreement to surrender, you will need to sign a simple declaration that you have received this notice and have accepted its consequences, before signing the agreement to surrender.

But if you do not receive at least 14 days notice, you will need to sign a "statutory" declaration. To do so, you will need to visit an independent solicitor (or someone else empowered to administer oaths).

Unless there is a special reason for committing yourself to the agreement to surrender sooner, you may want to ask the landlord to let you have at least 14 days to consider whether you wish to give up your statutory rights. If you then decided to go ahead with the agreement to end your lease, you would only need to make a simple declaration, and so you would not need to make a separate visit to an independent solicitor.

AND I make this solemn declaration conscientiously believing the same to be true and by virtue of the Statutory Declarations Act 1835

DECLARED at

---

this

---

day of

Before me (signature of person before whom declaration is made)

A commissioner for oaths or A solicitor empowered to administer oaths or (as appropriate)

Index

References are to paragraph numbers, not page numbers

For Product Safety Concerns and Information please contact our EU representative GPSR@taylorandfrancis.com Taylor & Francis Verlag GmbH, Kaufingerstraße 24, 80331 München, Germany

Batch number: 10732776

Printed by Printforce, the Netherlands